259
Topics in Current Chemistry

Topics in Current Chemistry

Recently Published and Forthcoming Volumes

Molecular Machines
Volume Editor: Kelly, T. R.
Vol. 262, 2006

Immobilisation of DNA on Chips II
Volume Editor: Wittmann, C.
Vol. 261, 2005

Immobilisation of DNA on Chips I
Volume Editor: Wittmann, C.
Vol. 260, 2005

Prebiotic Chemistry
From Simple Amphiphiles to Protocell Models
Volume Editor: Walde, P.
Vol. 259, 2005

Supramolecular Dye Chemistry
Volume Editor: Würthner, F.
Vol. 258, 2005

Molecular Wires
From Design to Properties
Volume Editor: De Cola, L.
Vol. 257, 2005

Low Molecular Mass Gelators
Design, Self-Assembly, Function
Volume Editor: Fages, F.
Vol. 256, 2005

Anion Sensing
Volume Editor: Stibor, I.
Vol. 255, 2005

Organic Solid State Reactions
Volume Editor: Toda, F.
Vol. 254, 2005

DNA Binders and Related Subjects
Volume Editors:Waring, M. J., Chaires, J. B.
Vol. 253, 2005

Contrast Agents III
Volume Editor: Krause,W.
Vol. 252, 2005

Chalcogenocarboxylic Acid Derivatives
Volume Editor: Kato, S.
Vol. 251, 2005

New Aspects in Phosphorus Chemistry V
Volume Editor: Majoral, J.-P.
Vol. 250, 2005

Templates in Chemistry II
Volume Editors: Schalley, C. A.,Vögtle, F., Dötz, K. H.
Vol. 249, 2005

Templates in Chemistry I
Volume Editors: Schalley, C. A.,Vögtle, F., Dötz, K. H.
Vol. 248, 2004

Collagen
Volume Editors: Brinckmann, J., Notbohm, H., Müller, P. K.
Vol. 247, 2005

New Techniques in Solid-State NMR
Volume Editor: Klinowski, J.
Vol. 246, 2005

Functional Molecular Nanostructures
Volume Editor: Schlüter, A. D.
Vol. 245, 2005

Natural Product Synthesis II
Volume Editor: Mulzer, J.
Vol. 244, 2005

Natural Product Synthesis I
Volume Editor: Mulzer, J.
Vol. 243, 2005

Prebiotic Chemistry

From Simple Amphiphiles to Protocell Models

Volume Editor: Peter Walde

With contributions by

L. Boiteau · A. Commeyras · G. Costanzo · C. Crestini · D. W. Deamer
E. DiMauro · J. P. Dworkin · C. Fernando · M. Lahav · L. Leiserowitz
R. Pascal · R. Saladino · M. Santos · E. Szathmáry · I. Weissbuch

Springer

The series *Topics in Current Chemistry* presents critical reviews of the present and future trends in polymer and biopolymer science including chemistry, physical chemistry, physics and material science. It is adressed to all scientists at universities and in industry who wish to keep abreast of advances in the topics covered.
As a rule, contributions are specially commissioned. The editors and publishers will, however, always be pleased to receive suggestions and supplementary information. Papers are accepted for *Topics in Current Chemistry* in English.
In references *Topics in Current Chemistry* is abbeviated *Top Curr Chem* and is cited as a journal.

Springer WWW home page: http://www.springeronline.com
Visit the TCC content at http://www.springerlink.com/

Library of Congress Control Number: 2005930232

ISSN 0340-1022
ISBN-10 3-540-27759-5 Springer Berlin Heidelberg New York
ISBN-13 978-3-540-27759-0 Springer Berlin Heidelberg New York
DOI 10.1007/b105128

Springer is a part of Springer Science+Business Media

springeronline.com

Printed in Germany

Cover design: *Design & Production* GmbH, Heidelberg
Typesetting and Production: LE-TEX Jelonek, Schmidt & Vöckler GbR, Leipzig

Printed on acid-free paper 02/3141 YL – 5 4 3 2 1 0

Volume Editor

Prof. Dr. Peter Walde

Department of Materials
Wolfgang-Pauli-Strasse 10
ETH Hoenggerberg, HCI F 501
8093 Zürich, Switzerland
peter.walde@mat.ethz.ch

Editorial Board

Topics in Current Chemistry Also Available Electronically

For all customers who have a standing order to Topics in Current Chemistry, we offer the electronic version via SpringerLink free of charge. Please contact your librarian who can receive a password or free access to the full articles by registering at:

springerlink.com

If you do not have a subscription, you can still view the tables of contents of the volumes and the abstract of each article by going to the SpringerLink Homepage, clicking on "Browse by Online Libraries", then "Chemical Sciences", and finally choose Topics in Current Chemistry.

You will find information about the

- Editorial Board
- Aims and Scope
- Instructions for Authors
- Sample Contribution

at springeronline.com using the search function.

Preface

Nobody knows how and where life originated. Although it is exceptionally difficult to define life and to define living systems, it is generally accepted that life as we know and experience it today on our planet is the result of biological (Darwinian) evolution. More and more complex living systems evolved from simpler ones. Prokaryotes are likely to be older than eukaryotes. Furthermore, it is generally assumed that all forms of life (current or extant) descend (or descended) from one or more first cell(s) that once emerged from the nonliving as a result of a prebiological chemical evolution on Earth or somewhere else. The first cells were probably preceded by some cell-like structures that were not yet cells but already resembled cells. These structures are generally called *protocells*, *protobionts* or *progenotes*.

Nobody knows what the first hypothetical cells and their precursor structures looked like. If one assumes that the first cells were already based on the three types of macromolecules that are so essential for the functioning of all contemporary cells – DNA, RNA, and proteins – then one has to admit that the first cells were already chemically very complex systems. As a logical consequence, it is assumed that protocells were simpler entities.

The field of prebiotic chemistry deals with experimental and theoretical approaches and scenarios toward an understanding of how biologically relevant molecules may have been formed and how they may have developed into protocellular systems, which eventually led to the origin of life. As a result of a number of studies, certain progress has been made during the last couple of years in at least five different areas that belong to the essential pieces of the big puzzle of understanding the origin of the first living systems. These five areas are covered in this special volume of *Topics in Current Chemistry*: (1) The prebiotic chemistry of simple amphiphilic molecules and their self-assembly into compartmentalized structures; (2) The prebiotic chemistry of nucleobases, nucleotides and oligonucleotides as essential elements of the hypothetical "RNA-world" and its relation to the proposed "pre-RNA-world(s)"; (3) The prebiotic chemistry of amino acids and peptides; (4) The possible prebiotic origin of homochirality; and (5) The possible formation of protocellular structures.

The chapters summarize a number of scientific hypothesis and recent experiments that are largely based on a variety of important contributions made by a large number of origin-of-life scientists and their research teams during the last century. Their experimental investigations, their ideas and their way of thinking have stimulated many researchers to study one of the most multidisciplinary and one of the most difficult question humans ever asked: the question of the origin of life, the the deepest roots of our admirable, miraculous living world.

Whether we will ever fully understand the transition from the nonliving to the living world, which is assumed to have happened at least about $3.5–3.8 \times 10^9$ years ago on our planet or somewhere else, is unclear at the moment. It is likely that we first need to make – in parallel to the developments in the field of prebiotic chemistry – further progress in a deeper, more detailed understanding of how living systems actually work. We need to better understand how the "elements" of all living systems, how cells function, how they are dynamically organized internally and how they communicate with the outside. It may well be that significant progress will be made during the next years within the field of systems biology that will further stimulate studies in the area of prebiotic chemistry. It is likely that completely new ideas and hypotheses about the origin of life will emerge as progress is also made in other research fields.

Zurich, July 2005 Peter Walde

Contents

Chemistry and Physics of Primitive Membranes
D. W. Deamer · J. P. Dworkin 1

On the Prebiotic Synthesis of Nucleobases, Nucleotides, Oligonucleotides, Pre-RNA and Pre-DNA Molecules
R. Saladino · C. Crestini · G. Costanzo · E. DiMauro 29

From the Prebiotic Synthesis of α-Amino Acids Towards a Primitive Translation Apparatus for the Synthesis of Peptides
R. Pascal · L. Boiteau · A. Commeyras 69

Stochastic "Mirror Symmetry Breaking" via Self-Assembly, Reactivity and Amplification of Chirality: Relevance to Abiotic Conditions
I. Weissbuch · L. Leiserowitz · M. Lahav 123

Evolutionary Potential and Requirements for Minimal Protocells
E. Szathmáry · M. Santos · C. Fernando 167

Author Index Volumes 251–259 213

Subject Index 219

Contents of Volume 253

DNA Binders and Related Subjects

Volume Editors: Michael J. Waring, Jonathan B. Chaires
ISBN: 3-540-22835-7

Regulation of Gene Expression by Synthetic DNA-Binding Ligands
P. B. Dervan · A. T. Poulin-Kerstien · E. J. Fechter · B. S. Edelson

Structural Selectivity of Drug-Nucleic Acid Interactions Probed by Competition Dialysis
J. B. Chaires

Cyanine Dye-DNA Interactions: Intercalation, Groove Binding and Aggregation
B. A. Armitage

Between Objectivity and Whim: Nucleic Acid Structural Biology
L. D. Williams

Topoisomerase Inhibitors of Marine Origin and Their Potential Use as Anticancer Agents
N. Dias · H. Vezin · A. Lansiaux · C. Bailly

DNA Major Groove Binders: Triple Helix-Forming Oligonucleotides, Triple Helix-Specific DNA Ligands and Cleaving Agents
C. Escudé · J.-S. Sun

Aminoglycoside-Nucleic Acid Interactions: The Case for Neomycin
D. P. Arya

Ribosomal RNA Recognition by Aminoglycoside Antibiotics
D. S. Pilch · M. Kaul · C. M. Barbieri

Top Curr Chem (2005) 259: 1–27
DOI 10.1007/b136806

Published online: 23 August 2005

Chemistry and Physics of Primitive Membranes

David W. Deamer[1] (✉) · Jason P. Dworkin[2]

[1]Department of Biomolecular Engineering, University of California, Santa Cruz, CA 95060, USA
deamer@hydrogen.ucsc.edu

[2]Laboratory for Astrochemistry, Solar System Exploration Division, NASA Goddard Space Flight Center, Code 691.0, Greenbelt, MD 20771, USA
Jason.P.Dworkin@nasa.gov

1 **Introduction** 2

2 **Self-Assembly Process in Early Forms of Life** 3

3 **Sources of Amphiphilic Compounds on the Early Earth** 4

4 **What Amphiphiles Composed the First Cell Membranes?** 8

5 **The Fluid Mosaic Model of Membrane Structure: Relation to Early Membranes** 9

6 **Function of Membranes in Early Cells** 10

7 **Growth Processes in Protocells** 13

8 **Encapsulation Mechanisms** 14

9 **Could Mineral-Water Interfaces Act as Precursors to Life?** 15

10 **Self-Assembly Processes in Prebiotic Organic Mixtures** 16

11 **Environmental Constraints on the First Cell Membranes** 19

12 **Model Systems of Primitive Cells** 20

13 **Summary** 24

References 24

Abstract A membrane boundary structure was essential for the advent of cellular life. The membranes of contemporary cells are composed of a mosaic of proteins embedded in a bimolecular layer of phospholipids, each of which requires a complex enzymatic pathway for its synthesis. The earliest forms of life could not have had such a highly evolved pathway in place. Amphiphilic monocarboxylic acids are present in carbonaceous meteorites and can be synthesized under simulated geochemical conditions. Such compounds have physical and chemical properties that allow them to assemble into bilayer membranes and are therefore plausible components of the first cellular membranes.

Keywords Artificial cells · Encapsulation · Lipid vesicles · Membranes

1 Introduction

Life on the Earth most likely arose from vast numbers of natural experiments in which various combinations of organic molecules were mixed and recombined to form complex interacting systems, then exposed to sources of energy such as light, heat, and oxidation-reduction potentials presented by donors and acceptors of electrons. This mixing and recombination probably did not occur in free solution, but rather in fluctuating environments at aqueous–mineral interfaces exposed to the atmosphere under conditions that would tend to concentrate the organic material so that reactions could occur. Through this process, incremental chemical evolution took place over a period of ten to several hundred million years after the Earth had cooled sufficiently for water vapor to condense into oceans. At some point, membrane-bounded systems of molecules appeared that could grow and reproduce by using energy and nutrients from the environment. An observer seeing this end product would conclude that such systems were alive but would be unable to pinpoint the exact time when the complex structures took on the property of life.

Here we will assume that the structures described above would be recognizable cells. The first cellular life had four key properties: (1) polymeric materials were encapsulated within a membrane-bounded structure; (2) the bounded system of molecules had the ability to capture energy and nutrients from the local environment; (3) the system could grow by spontaneous noncovalent addition of components from the environment and by catalyzed energy-dependent formation of covalent bonds between monomers to form polymers; and (4) the growing system could reproduce and evolve using a process directed by a replicating information-storage molecule.

Current research efforts have progressed to the point where the above processes have been investigated individually, so that the challenge now is to assemble them into an integrated system that exhibits the properties of the living state. This chapter focuses on the self-organizing properties of amphiphilic compounds that produce microscopic compartments necessary for the appearance of the first cellular forms of life.

2
Self-Assembly Process in Early Forms of Life

All cellular life today incorporates two processes we will refer to as self-assembly and directed assembly (Fig. 1). The latter involves the formation of covalent bonds by energy-dependent synthetic reactions and requires that a coded sequence in one type of polymer in some way direct the sequence of monomer addition in a second polymeric species. On the other hand, spontaneous self-assembly occurs when certain compounds associate through noncovalent hydrogen bonds, electrostatic forces, and nonpolar interactions that stabilize orderly arrangements of small and large molecules. Three well-known examples include the self-assembly of water molecules into ice, DNA

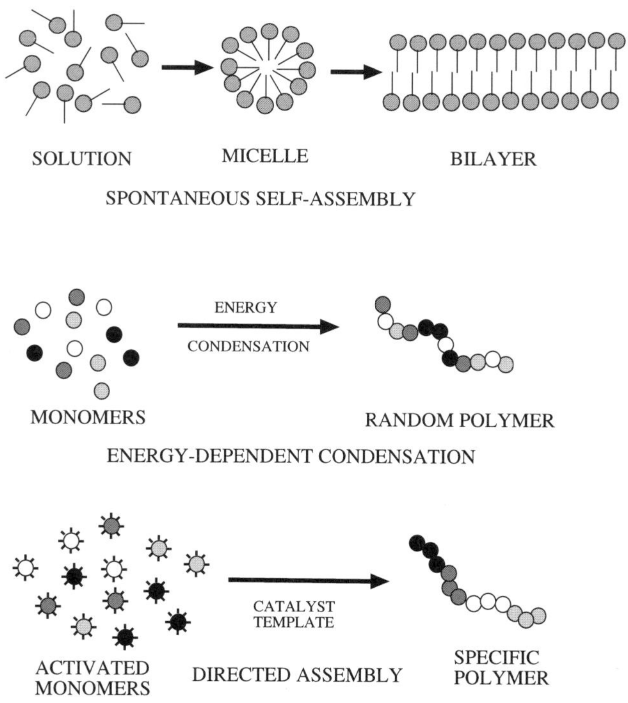

Fig. 1 Cellular life today uses both self-assembly and directed assembly processes to grow. Self-assembly (*upper diagram*) is essential to the synthesis and stability of membrane structures and protein folding, while directed assembly (*lower diagram*) underlies the synthesis of proteins according to the base sequences in DNA and mRNA. We assume that on the early Earth, random polymers similar to peptides and nucleic acids were produced by a yet unknown synthetic pathway (*center*). The random polymers, if capable of growth in a membrane-bounded microenvironment, would have been subjected to selection and thereby begin biological evolution

strands into a double helix, and newly synthesized protein chains into functional folded conformations. The latter two examples occur spontaneously, but the processes are enzyme mediated for regulatory reasons or to exclude undesirable conformations. A fourth self-assembly process involves certain compounds that can form closed membrane-bounded microenvironments. Such boundary structures, and the compartments they produce, have the potential to make energy available in the form of ion gradients and can provide a selective inward transport of nutrients. Furthermore, membranous compartments in principle are capable of containing unique systems of macromolecules. If a yet unknown macromolecular replicating system of polymers could be encapsulated within a membrane-bounded compartment, the components of the system would share the same microenvironment, and the result would be a major step toward cellularity, speciation, and true cellular function [1–6].

We know very little about how this event might have occurred at the origin of cellular life, but recent advances have provided clues about possible sources of amphiphilic molecules, assembly of membrane structures, and encapsulation mechanisms by which large molecules can be captured in membrane-bounded microenvironments. Here we will describe the chemical and physical properties of such systems and several experimental models that incorporate certain properties related to the origin of cellular life.

3
Sources of Amphiphilic Compounds on the Early Earth

There are only two possible sources of organic compounds on a primitive planetary surface: delivery during late accretion in the late Hadean era, followed by chemical evolution, or synthesis by geochemical processes in the primitive atmosphere and hydrosphere. Earlier investigations focused on chemical synthesis of monomers common to the primary macromolecules involved in living systems, with the goal of determining whether it was possible that biologically relevant compounds were available on the primitive Earth [7, 8]. Most of these studies emphasized water-soluble compounds such as amino acids, nucleobases, and simple carbohydrates. Here we will focus on self-assembling hydrocarbon derivatives. The most straightforward geochemical synthesis of hydrocarbons and their amphiphilic derivatives is the Fischer–Tropsch type synthesis (FTT). In this reaction, carbon monoxide is mixed with hydrogen and exposed to a hot catalyst such as metallic iron. Under these conditions, a remarkable reaction occurs in which hydrocarbon chains are synthesized by single-carbon additions, yielding alkanes, monocarboxylic acids, and alcohols ranging up to 30 or so carbons in length. Examples of such syntheses described in the literature include the pioneering observations of Oró et al. [10], with more recent results reported by McCollum et al. [11] and Rushdi et al. [12].

In the classic experiments of Miller and Urey [7, 8] the mixture of reduced gases was assumed to be a simulation of the original terrestrial atmosphere, which, by analogy with the outer planets, would have contained hydrogen, methane, ammonia, and water vapor. At sufficiently high energy fluxes, such mixtures of reduced gases generate hydrogen cyanide and formaldehyde, which in turn react by Strecker synthesis to produce amino acids, purines, and a variety of simple sugars. The proposal that organic compounds could be synthesized under prebiotic conditions was given additional weight when it was convincingly shown that carbonaceous meteorites contained amino acids, hydrocarbons, and even traces of purines [13–15]. If such meteorites represent samples of the primitive solar system components that underwent synthetic chemical reactions, it was reasonable to assume that similar reactions may have occurred on the Earth's surface.

This view was challenged in the late 1970s when lines of evidence emerged that the early atmosphere was composed of carbon dioxide and nitrogen rather than the mixture of reducing gases assumed by the Miller–Urey model [16, 17]. Carbon dioxide does not support synthetic pathways leading to chemical monomers [9], so interest was drawn to the second potential source of organic material: extraterrestrial infall in the form of micrometeorites and comets. This scenario was first proposed by Oró [18] and Delsemme [19] and more recently extended by Anders [20] and Chyba and Sagan [21]. The total organic carbon added by extraterrestrial infall over $\sim 10^8$ years of late accretion can be estimated to be in the rage of 10^{16}–10^{18} kg, which is several orders of magnitude greater than the total organic carbon in the biosphere. From such calculations it seems reasonable that extraterrestrial infall was a significant source of organic carbon in the prebiotic environment [22].

The discovery of biologically relevant compounds in meteorites also indicated that organic synthesis can occur in other environments, which immediately leads to the question of sources and synthetic pathways. Clues to a possible source of the meteoritic organics have been provided by infrared and millimeter astronomy. Vibrational and rotational spectral features obtained from molecular clouds indicate the presence of a plethora of carbon-containing compounds [23, 24]. Spectral features obtained from molecular clouds indicate the presence of a hundred or more carbon-containing compounds [23, 24]. Because dense molecular clouds are the birthplace of stars and solar systems, it seems reasonable that the organic substances present in comets and the parent bodies of meteorites were derived from the carbon compounds present in the original molecular cloud that gave rise to the solar system.

Dense molecular clouds attenuate the interstellar radiation field, permitting the synthesis and survival of more complex species in the gas phase than is possible in the diffuse interstellar medium. At the low temperatures in these dark molecular clouds (10–50 K), mixtures of molecules condense to form ice mantles on the surfaces of refractory dust grains where they can participate in additional gas-grain chemical reactions. Comparison of infrared spectra of low temperature laboratory ices with absorption spectra of molecular clouds indicates that interstellar ices are mainly composed of H_2O mixed with CO, CO_2, CH_3OH, NH_3, and other components, the latter ingredients generally comprising 5 to 15% of the total. The ices are exposed to ionizing radiation in the form of cosmic rays (and secondary radiation generated by their interaction with matter) and UV photons impinging upon the attenuated diffuse interstellar medium (ISM) or generated by stars forming within the cloud.

Laboratory experiments have shown that radiation processing of simulated presolar ices leads to more complex molecular species [25–27]. Hundreds of new compounds are synthesized, although the starting ices contain only a few simple common interstellar molecules. Many of the compounds formed in these experiments are also present in meteorites and cometary and asteroidal dust (interplanetary dust particles – IDPs), and some are presumably relevant to the origin of life, including amino acids [28, 29], quinines [30], and amphiphilic material [31].

The consensus view is that organic molecules and their building blocks are synthesized in dense molecular clouds and then become components of the presolar nebula where they are further altered; these nebula give rise to stars, solar systems, and the parent bodies of meteorites. However, the molecules must be delivered to habitable planetary surfaces if they are to take part in the origin of life. This requires that they survive the transition from the dense cloud into a protostellar nebula and subsequent incorporation into planetesimals, followed by delivery to a planetary surface. Theoretical calculations suggest that a fraction of the extraterrestrial organics present in comets should survive even during impact with a planetary atmosphere [32], and experimental results confirm that some organic species do, in fact, survive planetary accretion. The most convincing evidence comes from deuterium isotopic measurements of meteorites and interplanetary dust particles (IDPs) collected on Earth. Such objects contain many of the same compounds and classes of compounds produced in interstellar simulations, and meteoritic organics frequently have large deuterium excesses [33]. These excesses are difficult to understand in terms of solar system chemistry but may be explained by a variety of interstellar chemical processes that produce organic compounds [34].

Even today, meteorites and IDPs deliver organic materials to the Earth's surface at a rate of $\sim 1 \times 10^6$ kg/year [35]. During the late bombardment period, which lasted until about 4 billion years ago, the amount of extraterrestrial organic material brought to the prebiotic Earth was likely to have been orders of magnitude greater [21]. Thus, the early Earth must have been seeded with organic matter created in the interstellar medium, protosolar nebula, and asteroidal/cometary parent bodies.

From these considerations we conclude that both exogenous delivery and endogenous synthetic pathways provided organic material to the prebiotic environment, a process summarized in Fig. 2. We cannot be certain about the relative amounts of endogenous synthesis and extraterrestrial delivery of organics, but it is likely that both sources played significant roles in the subsequent emergence of life by providing specific molecular species that were essential ingredients. The question to be answered now is what fraction was degraded to simple carbon compounds such as CO and CO_2 and what fraction was incorporated directly into the molecular systems leading to the origin of life.

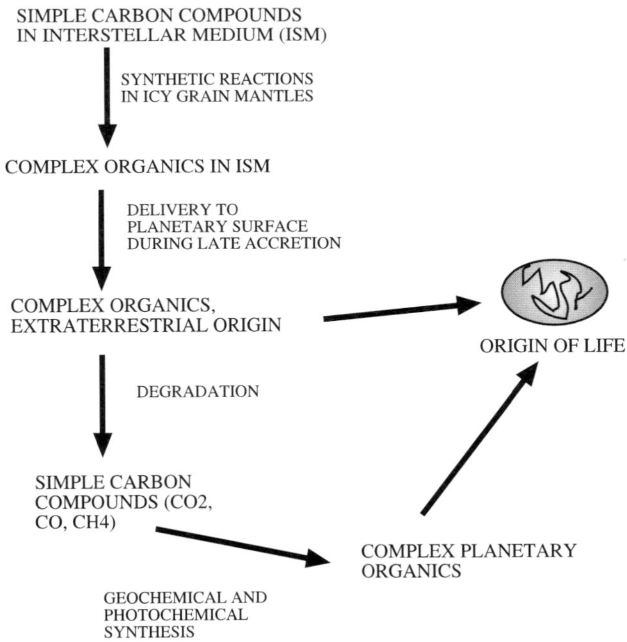

Fig. 2 All terrestrial carbon was initially delivered to the primitive Earth during accretion. Much of the carbon was degraded to simple carbon compounds that could then undergo synthetic geochemical reactions to produce more-complex species. However, a fraction of the delivered organic carbon was likely to survive intact, especially during late accretion. This fraction had the potential to be incorporated into the molecular systems that gave rise to the origin of life

4
What Amphiphiles Composed the First Cell Membranes?

Amphiphilic molecules are among the simplest of life's molecular components and are readily synthesized by nonbiological processes. Virtually any normal alkane having ten or more carbons in its chain takes on amphiphilic properties if one end of the molecule incorporates a polar or ionic group (see below). The simplest common amphiphiles are therefore molecules such as monocarboxylic acids (anions), monoamines (cations), and alcohols (neutral polar groups).

$$CH_3-(CH_2)_n-COOH \rightarrow H^+ + CH_3-(CH_2)_n-COO^- \quad \text{(anion)}$$
$$CH_3-(CH_2)_n-NH_2 + H^+ \rightarrow CH_3-(CH_2)_n-NH_3^+ \quad \text{(cation)}$$
$$CH_3-(CH_2)_n-OH \quad \text{(neutral amphiphile)}$$

Lipids are far more diverse chemically than other typical biomolecules such as amino acids, carbohydrates, and nucleotides. The definition of lipids includes simple fatty acids and their glycerol esters, sterols such as cholesterol, and phospholipids, sphingolipids, and cerebrosides. Lipids are generally defined by their common hydrophobic character, which makes them soluble in organic solvents such as chloroform. Virtually all lipids also have a hydrophilic group, which makes them surface active.

Eukaryotic phospholipids typically have two fatty acid chains linked to a glycerol by ester bonds, with the third position of the glycerol esterified to a phosphate group. Most phospholipids also have a head group such as choline, ethanolamine, or serine attached to the phosphate, and one such lipid is shown below (1-palmitoyl, 2-oleoyl phosphatidylcholine). The precise function of the variable head groups has not yet been established.

Scheme 1

The other lipid commonly present in eukaryotic membranes is cholesterol, a polycyclic structure produced from isoprene by a complex biosynthetic pathway. It is interesting to ask whether it is conceivable that prebiotically plausible reactions might also produce complex amphiphiles. The earliest investigations aiming to answer this question were carried out by Hargreaves et al. [36], Oró and coworkers [37, 38], and, more recently, Ourisson et al. [39] and Conde-Frieboes and Blochliger [40]. In all such reactions, energy-dependent condensation reactions are used to produce complex lipids from mixtures of phosphate, fatty acid, and glycerol. Examples of such lipids include phosphatidic acid and phosphatidylcholine, both of which readily self-assemble into membranous vesicles.

Although it is clear that complex lipids can be synthesized under laboratory simulations using pure reagents, the list of required ingredients does not seem plausible under prebiotic conditions. Therefore, it is unlikely that early membranes were composed of complex lipids such as phospholipids and cholesterol. Instead, there must have been a source of simpler amphiphilic molecules capable of self-assembly into membranes. One possibility is lipid-like fatty acids and fatty alcohols, which are products of FTT simulations of prebiotic geochemistry [12] and are also present in carbonaceous meteorites. Furthermore, as will be discussed later, these compounds form reasonably stable lipid bilayer membranes by self-assembly from mixtures (Fig. 4a).

5 The Fluid Mosaic Model of Membrane Structure: Relation to Early Membranes

In the 1970s, the fluid mosaic concept emerged as the most plausible model to account for the known structure and properties of biological membranes [41]. The fact that membranes exist as two-dimensional fluids (liquid disordered) rather than in a gel state (solid ordered) was clearly demonstrated by Frye and Ededin [42], who showed that the lipid and protein components of two separate membranes diffuse into each other when two different cells were fused. Since that time, numerous studies have measured the diffusion coefficient of lipids and proteins in membranes, and the diffusion rates were found to correspond to those expected of a fluid with the viscosity of olive oil rather than a gel phase resembling wax.

Because the lipid components of membranes must be in a fluid state to function as membranes in living cells, it is reasonable to assume that primitive membranes in the first forms of cellular life must also have had this property. Straight-chain hydrocarbons have relatively high melting points due to the ease with which van der Waals interactions can occur along the chains. Any discontinuity in the chains interrupts these interactions and markedly decreases the melting point. As an example, stearic acid contains 18 carbons in its alkane chain and melts at 68 °C, while oleic acid, with a cis-double bond between carbons 9 and 10, has a melting point near 14 °C. If cellular life today requires fluid membranes, it is reasonable to assume that the earliest cell membranes were also composed of amphiphilic molecules in a fluid state.

The idea that the proteins of biological membranes are embedded in a fluid sea of lipids arose from our increasing understanding of membrane structure. It has been demonstrated in numerous ways that most of the proteins associated with membranes are embedded in the lipid bilayer phase, rather than simply adhering to the surface. As a general rule, membrane proteins have stretches of hydrophobic amino acids in their sequences, and these are threaded back and forth through the bilayer multiple times, thereby anchoring the protein to the membrane. The hydrophobic proteins often are involved in production of pores, or transmembrane channels, that are essential for ion and nutrient transport processes.

Could similar channels be produced in the bilayer membranes of primitive cells? There is no doubt that channel-like defects appear when a nonpolar peptide interacts with a lipid bilayer. For instance, polyleucine or polyalanine has been induced to fuse with planar lipid membranes, and the bilayers exhibited transient bursts of proton conductance [43]. Surprisingly, channel-like conductance also appears when RNA is selected for its ability to bind to phospholipids [44]. From these observations it is fair to say that if random polymers were being produced by some unknown synthetic reaction on the early Earth, some of those polymers were likely to have been able to penetrate bilayer membranes and produce channels that bypassed the permeability barrier. This is an area that is ripe for further investigations, as described in a recent review by Pohorille et al. [45].

6
Function of Membranes in Early Cells

Membranes have many functions in addition to acting as a container for the macromolecular polymers of life. Three primary membrane functions associated with a protocell would include selective inward transport of nutrients from the environment, capture of the energy available in light or oxidation-reduction potentials, and coupling of that energy to some form of energy currency such as ATP in order to drive polymer synthesis (Fig. 3).

The simplest of these functions is that of a permeability barrier that limits free diffusion of solutes between the cytoplasm and external environment. Although such barriers are essential for cellular life to exist, there must also be a mechanism by which selective permeation allows specific solutes to cross the membrane. In contemporary cells, such processes are carried out by transmembrane proteins that act as channels and transporters. Examples include the proteins that facilitate the transport of glucose and amino acids into the cell, channels that allow potassium and sodium ions to permeate the membrane, and active transport of ions by enzymes that use ATP as an energy source.

It seems unlikely that the first living cellular systems had time to evolve highly specialized membrane transport systems, which brings up the ques-

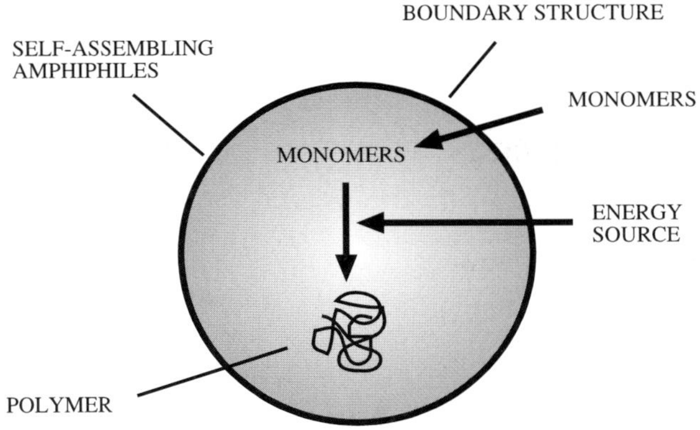

Fig. 3 A protocell would have had a minimal set of functional properties, including self-assembly of boundary membranes, transport of monomers, and capture of energy to drive polymerization reactions, and encapsulation of polymer systems capable of growth

tion of how early cells overcame the membrane permeability barrier. One possibility is that simple diffusion across the bilayer may have been sufficient. To give a perspective on permeability and transport rates by diffusion, we can compare the fluxes of relatively permeable and relatively impermeable solutes across contemporary lipid bilayers. The measured permeability of lipid bilayers to small, uncharged molecules such as water, oxygen, and carbon dioxide is greater than the permeability to ions by a factor of $\sim 10^9$. For instance, the permeability coefficient of water is approximately 10^{-3} cm/s, and the permeability coefficient of potassium ions is 10^{-11} cm/s. These values mean little by themselves, but make more sense when put in the context of time required for exchange across a bilayer. Measurements show that half the water in a liposome exchanges in milliseconds, while potassium ions have half-times of exchange measured in days.

We can now consider some typical nutrient solutes like amino acids and phosphate. Such molecules are ionized, which means that they would not readily cross the permeability barrier of a lipid bilayer. Permeability coefficients of liposome membranes to phosphate and amino acids have been determined [46] and were found to be in the range of 10^{-11}–10^{-12} cm/s, similar to ionic solutes such as sodium and chloride ions. From these figures one can estimate that if a primitive microorganism depended on passive transport of phosphate across a lipid bilayer composed of a typical phospholipid, it would require several years to accumulate phosphate sufficient to double its DNA content or pass through one cell cycle. In contrast, a modern bacterial cell can reproduce in as short a time as 20 min.

If bilayers are so impermeable to solutes like amino acids and phosphate, how could primitive cells have had access to these essential nutrients? One clue may be that modern lipids are highly evolved products of several billion years of evolution and typically contain hydrocarbon chains 16 to 18 carbons in length. These chains provide an interior "oily" portion of the lipid bilayer that represents a nearly impermeable barrier to the free diffusion of ions such as sodium and potassium. The reason is related to the common observation that "oil and water don't mix," That is, ion permeation of the hydrophobic portion of a lipid bilayer faces a very high energy barrier called Born energy, which is associated with the difference in energy for an ion in a high dielectric medium (water with a dielectric constant of 80) compared to the same ion in a low dielectric medium (hydrocarbon with a dielectric constant of 2). This energy barrier is immense, up to 40 kcal/mole [47].

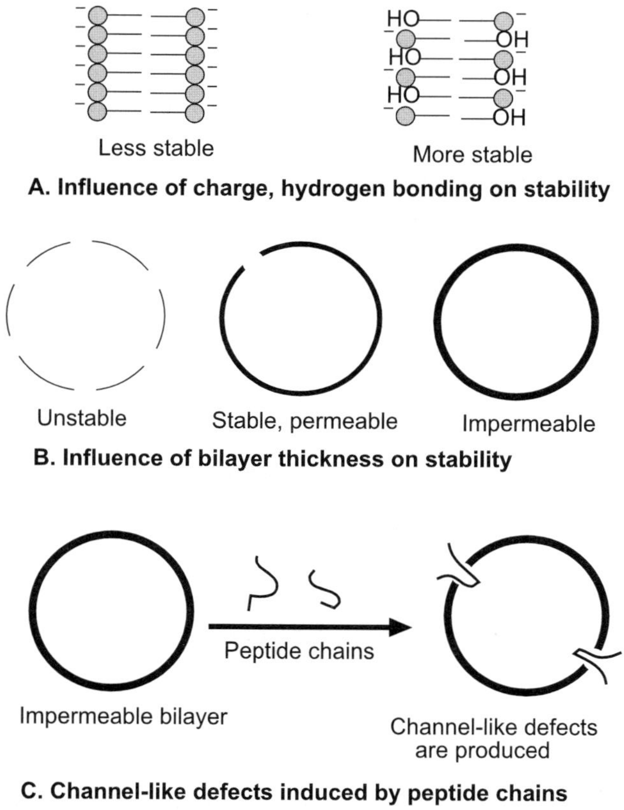

Fig. 4 Stability and permeability of self-assembled amphiphilic structures. Amphiphilic molecules such as fatty acids having carbon chain lengths of 9 or more carbons form bilayer membranes when sufficiently concentrated. **a** Pure bilayers of ionized fatty acid are relatively unstable but become markedly more stable as long chain alcohols are added. **b** Dimensions of the amphiphile also play a role. Shorter chain amphiphiles (9–10 carbons) are less able to form bilayers, while those of intermediate chain length (12–14 carbons) produce stable bilayers that also are permeable to ionic and polar solutes. Longer chain lengths (16–18 carbons) produce bilayers that are increasingly less permeable to solutes [48]

However, recent studies have shown that permeability is strongly dependent on chain length [48]. For instance, shortening phospholipid chains from 18 to 14 carbons increases permeability to ions by a thousandfold (Fig. 4b). The reason is that thinner membranes have increasing numbers of transient defects that open and close on nanosecond time scales, so that ionic solutes can get from one side of the membrane to the other without dissolving in the oily interior phase of the bilayer. Ionic solutes even as large as ATP can diffuse across a bilayer composed of dimyristoylphosphatidylcholine, a 14-carbon phospholipid [49]. On the early Earth, shorter hydrocarbon chains would have been much more common than longer chain amphiphiles, suggesting that the first cell membranes were sufficiently leaky so that ionic and polar nutrients could enter while still maintaining larger polymeric molecules in the encapsulated volume. However, it should be kept in mind that polymeric compounds may also have been present in the prebiotic environment. If these happened to interact with the lipid bilayer of a membrane compartment, channel-like defects could have been produced that permitted transport of polar and ionic nutrients (Fig. 4c).

7 Growth Processes in Protocells

Earlier reports [50] showed that vesicles composed of oleic acid can grow and "reproduce" as oleoyl anhydride spontaneously hydrolyzed in the reaction mixture, thereby adding additional amphiphilic components (oleic acid) to the vesicle membranes. This approach has recently been extended by Hanczyc et al. [51], who prepared myristoleic acid membranes under defined conditions of pH, temperature, and ionic strength. The process by which the vesicles formed from micellar solutions required several hours, apparently with a rate-limiting step related to the assembly of "nuclei" of bilayer structures. However, if a mineral surface in the form of clay particles was present, the surface in some way catalyzed vesicle formation, reducing the time required from hours to a few minutes. The clay particles were spontaneously encapsulated in the vesicles. The authors further found that RNA bound to the clay was encapsulated as well.

In a second series of experiments, Hanczyc et al. [52] found that the myristoleic acid vesicles could be induced to grow by addition of fatty acid to the medium, presumably by incorporating fatty acid molecules into the membrane, rather than by fusion of vesicles. If the resulting suspension of large vesicles was then filtered through a polycarbonate filter having pores 0.2 μm in diameter, the larger vesicles underwent a kind of shear-induced division to produce smaller vesicles. This process could be repeated several times (Fig. 5).

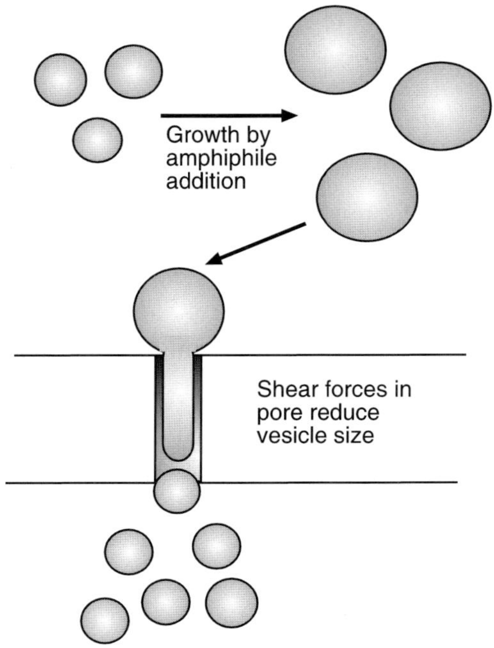

Fig. 5 Fatty acid vesicles can grow by addition of fatty acid molecules to the membrane and are then dispersed into smaller vesicles by passage through a porous filter. The cycle of growth and dispersion was repeated several times and presumably could go on indefinitely [52]

This remarkable series of experiments clearly demonstrated the relative simplicity of producing a complex system of lipid, genetic material, and mineral catalyst in a model protocellular structure that can undergo a form of growth and division.

8
Encapsulation Mechanisms

Even if membranous vesicles were commonplace on the early Earth and had sufficient permeability to permit nutrient transport to occur, these structures would be virtually impermeable to larger polymeric molecules that were necessarily incorporated into molecular systems on the pathway to cellular life. The encapsulation of macromolecules in lipid vesicles has been demonstrated by hydration–dehydration cycles that simulate an evaporating lagoon [53] or by freeze–thaw cycles [54]. Molecules as large as DNA can be captured by such processes. For instance, when a dispersion of DNA and fatty acid vesicles is dried, the vesicles fuse to form a multilamellar sandwich structure with

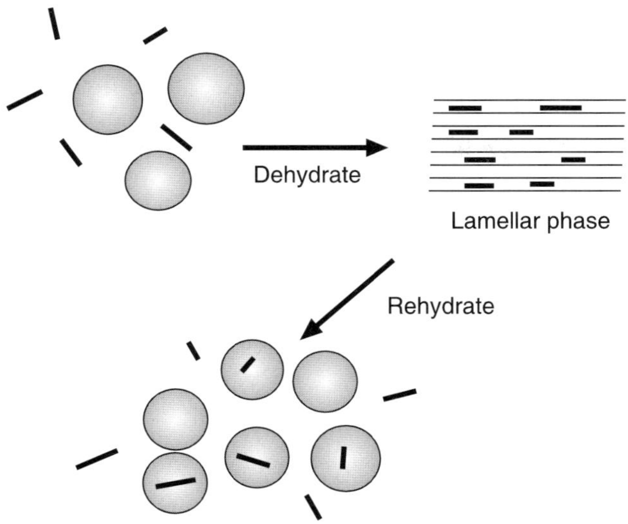

Fig. 6 Macromolecules are readily encapsulated in lipid vesicles in a single cycle of dehydration–hydration [53]. Such wetting–drying cycles would have commonly occurred in the prebiotic environment at intertidal zones

DNA trapped between the layers. Upon rehydration, vesicles reform that contain highly concentrated DNA, a process that can be visualized by staining with a fluorescent dye (Fig. 6). Several enzymes have also been encapsulated using similar procedures [55].

9
Could Mineral-Water Interfaces Act as Precursors to Life?

Given that organic compounds are made available on a planetary surface as described above, they must then be organized into mixtures that are sufficiently concentrated to undergo chemical reactions. It seems likely that most of the compounds would be delivered to extensive early oceans, rather than the relatively small area of available volcanic terrain, but this immediately leads to a conundrum: global concentrations of organic material in the seas would have been too dilute to undergo synthetic chemical reactions [56]. For this reason, it has been suggested that mineral/aqueous interfaces were primary agents in concentrating and organizing organic solutes, and perhaps catalyzing specific reactions related to life processes [57, 58]. Corliss et al. [59] and Baross and Hoffman [60] first proposed that life began as an organic film on mineral surfaces in subsurface geothermally active sites. Such films would provide a microenvironment of low water activity so that hydrolytic back reactions would not continuously degrade more complex molecules such as polymers formed by condensation reactions. As an energy source, either dissolved hydrogen gas or the mineral surface itself would provide a source of reducing power [61–63]. In this scenario, membrane encapsulation and a system of information transfer would evolve at some later time.

As a specific chemical example of a mineral-dependent reaction pathway, Huber and Wächtershäuser [64] described an experimental model in which a slurry of nickel and iron sulfide was found to promote the formation of acetic acid from carbon monoxide and methyl mercaptan (CH_3SH). Peptide bond formation could also be demonstrated [65]. These conditions were considered to represent a simulation of a primordial geothermal system in which metal sulfides at high temperatures ($\sim 100\,^{\circ}C$) provide a reaction pathway for the initial steps of an autotrophic metabolism. This is an interesting result that is pertinent to the synthesis of organic material and confirms earlier observations that a variety of free energy sources can drive the formation of simple organic molecules. In the case of a mineral surface, the chemical potential is contained in the reactants and the mineral itself, rather than an impinging energy source such as electrical discharge or UV photons.

Martin and Russell [66] have taken this concept a step further. They note that certain iron sulfide minerals contain microscopic pores in the size range of cells (~ 10–$100\,\mu m$) and propose that such cavities could provide a mineral version of a membranous boundary structure. The authors suggest that the cavities may be able to concentrate nutrient organic solutes that could serve as reactants in primitive metabolic pathways. They also propose that the iron sulfide membranes could provide a source of chemical energy, perhaps even chemiosmotic energy to drive early metabolism. This idea has merit and deserves further testing. In particular, it should be determined whether mineral membranes are able to act as true permeability barriers to the free diffusion of solutes. So far, permeability barriers have only been demonstrated in lipid bilayer membranes having a hydrophobic phase that has the capacity to maintain concentration gradients of polar and ionic solutes.

10
Self-Assembly Processes in Prebiotic Organic Mixtures

What physical properties are required if a molecule is to become incorporated into a stable bilayer? As discussed earlier, all bilayer-forming molecules are amphiphiles, with a hydrophilic "head" and a hydrophobic "tail" on the same molecule. If amphiphilic molecules were present in the mixture of organic compounds available on the early Earth, it is not difficult to imagine that their self-assembly into molecular aggregates was a common process.

Is this a plausible premise? In order to approach this question, we can assume that the mixture of organic compounds in carbonaceous meteorites such as the Murchison meteorite resembles components available on the early Earth through extraterrestrial infall. A series of organic acids represents the most abundant water-soluble fraction in carbonaceous meteorites [15, 67, 68]. Samples of the Murchison meteorite were extracted in an organic solvent commonly used to extract membrane lipids from biological sources [69, 70]. When this material was allowed to interact with aqueous phases, one class of compounds with acidic properties was clearly capable of forming membrane-bounded vesicles (Fig. 7).

Significantly, the photoproducts of interstellar ice simulations also include amphiphilic compounds having self-assembly properties [31]. Figure 8 shows micrographs of Murchison vesicles, as well as vesicles formed by products of interstellar ice simulations and known fatty acid–fatty alcohol mixtures. It is clear that the vesicle-forming behavior of all of these amphiphiles is

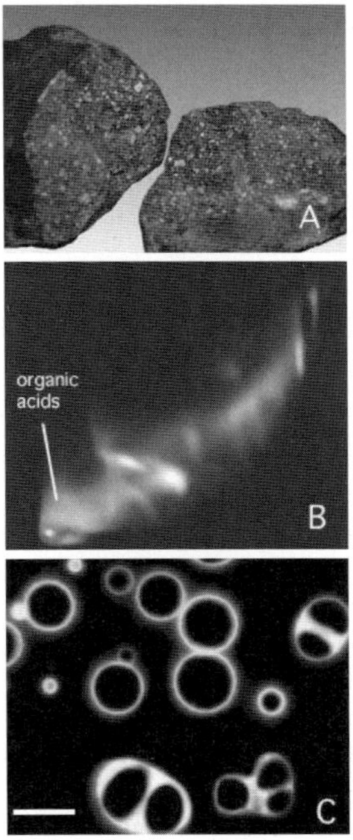

Fig. 7 Membranes can be formed by components of carbonaceous meteorites [69, 70]. **a** The Murchison meteorite contains approximately 2% organic carbon by weight. **b** Organic compounds can be extracted from the meteorite by a lipid solvent system (chloroform–methanol), then separated by two-dimensional chromatography. Polycyclic compounds in the mixture produce fluorescent spots. **c** The organic acid fraction from the TLC plate readily assembles into membranous vesicles when exposed to dilute aqueous solutions buffered at pH 8–9. The vesicles were photographed by their autofluorescence. *Scale bar* shows 20 μm

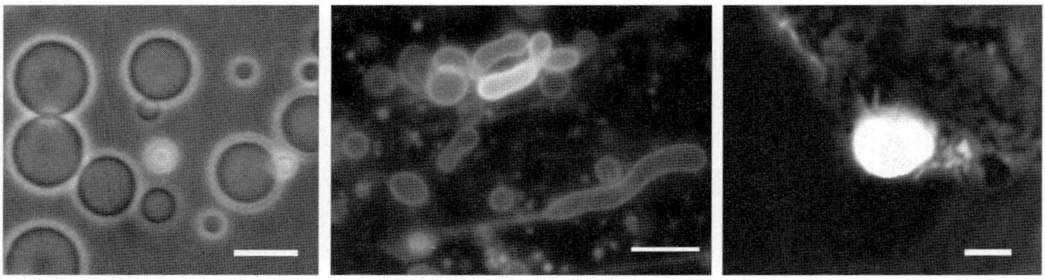

Fig. 8 Phase and fluorescence micrographs of membranous vesicular structures formed from a Murchison meteorite extract (*left*) compared to vesicles formed by a 20 mM decanoic acid–decanol mixture [72] (*center*) and a vesicular structure produced by the photoproduct of an interstellar-ice analog [31]. The vesicles produced by the photochemical ice analog product were allowed to capture pyranine, a fluorescent anionic dye, to demonstrate that a true membrane was present. *Scale bars* show 20, 10, and 5 μm, from *left* to *right*

similar. The organic compounds present in the meteoritic extract and those synthesized in the simulation of grain mantle photochemistry both contain amphiphilic compounds capable of self-assembly into membranous boundary structures. The vesicles produced from the interstellar simulations, like those of the meteoritic compounds, can also capture and maintain a gradient of ionic dye molecules.

From these results it is reasonable to conclude that a variety of simpler amphiphilic molecules were present on the early Earth that could participate in the formation of primitive membrane structures. The long chain acids and alcohols that contribute the amphiphilic property of contemporary membrane lipids are one possible component of prebiotic membrane structures. These compounds are present in carbonaceous meteorites [67, 68] and are synthesized under simulated geochemical conditions [11, 12]. Significantly, such simple amphiphiles can also form vesicles, as shown earlier [71, 72]. Stability of the vesicles is strongly dependent on chain length, concentration, amphiphile composition, temperature, and head group characteristics. For example, even a 9-carbon monocarboxylic acid—nonanoic acid—can form vesicles at concentrations of 85 mM and pH 7.0, which is the pK of the acid in bilayers [70]. Addition of small amounts of an alcohol (nonanol) further stabilizes the bilayers due to hydrogen bonding between the alcohol and acid head groups, so that vesicles can form at lower concentrations ($\sim$ 20 mM) at pH ranging from 6 to 11 (Fig. 4a). The vesicles provide a selective permeability barrier, as indicated by osmotic activity and ionic dye capture. As chain length increases, stability also increases and vesicles form at lower concentrations.

11
Environmental Constraints on the First Cell Membranes

Although self-assembly of amphiphilic molecules promotes the formation of complex molecular systems, the physical and chemical properties of an aqueous phase can significantly inhibit such processes, possibly constraining the environments in which cellular life first appeared. One such constraint is that temperature strongly influences the stability of vesicle membranes. It has been proposed that the last common ancestor, and even the first forms of life, were hyperthermophiles that developed in geothermal regions such as hydrothermal vents [60] or deep subterranean hot aquifers [61]. Such environments have the advantage of providing chemical energy in the form of redox potentials as well as abundant mineral surfaces to act as potential catalysts and adsorbants. However, because the intermolecular forces that stabilize self-assembled molecular systems are relatively weak, it is difficult to imagine how lipid bilayer membranes assembling from plausible prebiotic constituents would be stable under these conditions. All hyperthermophiles today have highly specialized lipid components, and it seems likely that these are the result of more recent adaptation than a molecular fossil of early life.

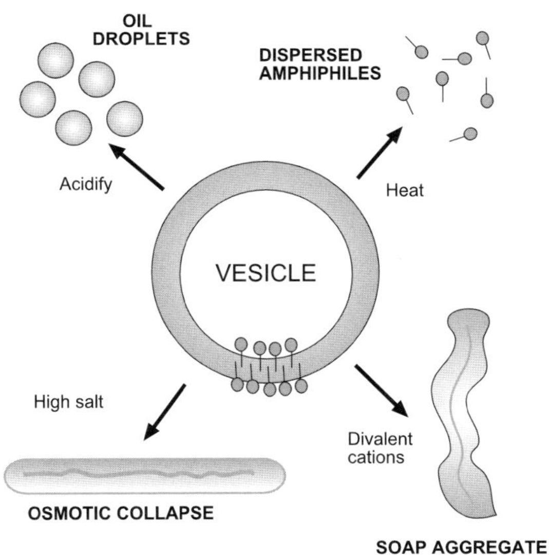

Fig. 9 Vesicles produced by single-chain amphiphiles such as fatty acids tend to be destabilized by certain environmental factors. If the fatty acid is protonated at low pH ranges, the membranes collapse into droplets. The vesicles also become increasingly unstable as temperature increases. In the presence of high salt concentrations, the vesicles undergo osmotic collapse and may also form nonmembranous aggregates if divalent cations react with the carboxylate head groups

A second concern is related to the ionic composition of a marine environment. The high salt concentration of the present ocean (near 0.5 M NaCl) has the potential to exert significant osmotic pressure on any closed membrane system (Fig. 9). All marine organisms today have evolved highly developed membrane transport systems that allow them to maintain osmotic equilibrium against substantial salt gradients across their membranes. Furthermore, the concentrations of divalent cations, in particular Mg^{2+} and Ca^{2+}, were likely to exceed 10 mM in the early oceans. In the absence of oxygen, Fe^{2+} would also be present at similar concentrations. All such divalent cations have a strong tendency to bind to the anionic head groups of amphiphilic molecules, strongly inhibiting their ability to form stable membranes [73].

These considerations suggest that, from the perspective of membrane biophysics, the most plausible planetary environment for the origin of life would be at moderate temperature ranges (< 60 °C), and the ionic content would correspond to low ionic strength and pH values near neutrality (pH 5–8) with divalent cations at submillimolar concentrations. This suggestion is in marked contrast to the view that life most likely began in a marine environment, perhaps even the extreme environment of a hydrothermal vent. A marine site for life's beginning seems plausible because freshwater would be rare on the early Earth. Even with today's extensive continental crust, freshwater only represents $\sim$ 1% of the contemporary Earth's reservoir of liquid water. Another concern about a freshwater origin of life is that the lifetime of freshwater bodies tends to be geologically short-lived. On the other hand, if seawater, with its high content of sodium chloride and divalent ions, markedly inhibits self-assembly processes and reactions that are essential to the emergence of cellular life, we may need to reconsider the assumption that life inevitably began in a marine environment. A more plausible site for the origin of cellular life may be a low-ionic-strength lacustrine environment such as a pond or lake. After the first form of cellular life was able to establish itself in a relatively benign environment, it would rapidly begin to adapt through Darwinian selection to more rigorous environments, including the extreme temperatures, salt concentrations, and pH ranges that we associate with the limits of life on the Earth.

12
Model Systems of Primitive Cells

A central event in the origin of life was the self-assembly of a molecular system in which catalytic polymers could interact with a second class of polymers having the capacity to store information in a sequence of monomers. That sequence in turn would in some manner determine the sequence of monomers in the catalyst, so that the resulting catalytic-information cycle was able to undergo directed growth. In contemporary cells, the cycle is represented by protein catalysts (enzymes) and nucleic acids that store genetic information and have the potential to transmit that information to a second molecule by replication or transcription. However, in a protocell, both catalytic and information-containing sites could be present in the same molecule, as suggested by recent studies of RNA ribozymes [74]. Several approaches to artificial cells have been proposed to test various scenarios for the origin of cellular life [75–78]. An ideal model cell would incorporate an encapsulated polymerase activity together with a template of some sort, so that sequence information in the template could be transcribed to a second molecule. The membrane must be sufficiently permeable to allow the polymerase to have access to externally added substrates. Furthermore, the membrane itself should be able to grow in order to accommodate the growth of the encapsulated polymers. Finally, in an ideal cell model, the polymerase itself would be reproduced from information in the template, so that the entire system would be able to grow and evolve.

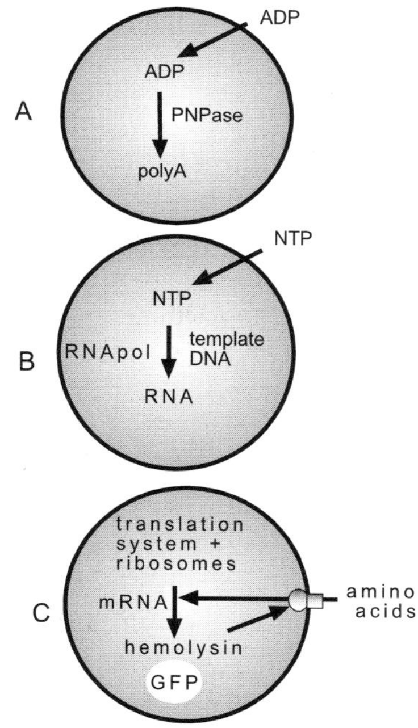

Fig. 10 Model protocell systems. **a** An encapsulated polymerase (polynucleotide phosphorylase) can synthesize RNA from nucleoside diphosphates such as ADP [79, 80]. **b** RNA can be synthesized by a template-dependent T7 RNA polymerase [83]. **c** Proteins such as green fluorescent protein (GFP) can be synthesized by an encapsulated translation system [84]. If mRNA coding for hemolysin is also present, the hemolysin forms a pore in the lipid bilayer. Amino acids then permeate the bilayer, and protein synthesis can continue for several days [85]

To demonstrate polymerase activity in a model cell, Chakrabarti et al. [79] encapsulated polynucleotide phosphorylase in vesicles composed of dimyristoylphosphatidylcholine (DMPC). This enzyme can produce RNA from nucleoside diphosphates such as adenosine diphosphate (ADP) and does not require a template, so it has proven useful for initial studies of encapsulated polymerase activity (Fig. 10a). Furthermore, DMPC liposomes are sufficiently permeable so that 5–10 ADP molecules per second enter each vesicle. Under these conditions, measurable amounts of RNA in the form of polyadenylic acid were synthesized and accumulated in the vesicles after several days' incubation. The enzyme-catalyzed reaction could be carried out in the presence of a protease external to the membrane, demonstrating that the vesicle membrane protected the encapsulated enzyme from hydrolytic degradation. Similar behavior has been observed with monocarboxylic acid vesicles [80], and it follows that complex phospholipids are not required for an encapsulated polymerase system to function.

In other work, the Q-beta replicase [81] and the components of the polymerase chain reaction (PCR) [82] have also been encapsulated, together with templates and substrates in the form of nucleoside triphosphates (NTPs), and are functional in liposomes. Both of these enzyme systems use templates, so it is clear that template-dependent polymer synthesis can occur in an encapsulated environment. The phospholipids used in these studies were relatively impermeable, so substrates were necessarily encapsulated along with enzyme and template. This limited the amount of nucleic acid replication that could occur to a few molecules per vesicle. However, the permeability barrier can in principle be overcome by introducing transient defects in the membranes of lipid vesicles. For instance, a template-directed reaction can be encapsulated in DMPC liposomes in which externally added substrates were used to supply the enzyme [83]. In this study, T-7 RNA polymerase and a circular 4000 bp plasmid template were encapsulated, and substrates were provided by addition of the NTPs (Fig. 10b). The system was subjected to temperature cycles of 23 °C and 37 °C in a PCR apparatus. DMPC membranes are relatively permeable at the phase transition temperature of 23 °C, permitting substrate ribonucleotides to enter the vesicles, while at 37 °C the membranes become much less permeable but the polymerase is activated. RNA synthesis was monitored by incorporation of radiolabeled UTP, and transcription was confirmed by reverse PCR. Figure 11 shows a micrograph of the resulting structures containing RNA synthesized within the vesicle volume.

Most recently, functioning translation systems that included ribosomes have been encapsulated in lipid vesicles. The first attempt to assemble a translation system in a lipid vesicle system was made by Oberholzer et al. [84]. However, only very small amounts of peptides were synthesized, largely because the lipid bilayer was impermeable to amino acids, so that ribosomal translation was limited to the small number of amino acids encapsulated within the vesicles. Yu et al. [85] and Nomura et al. [86] improved the yield

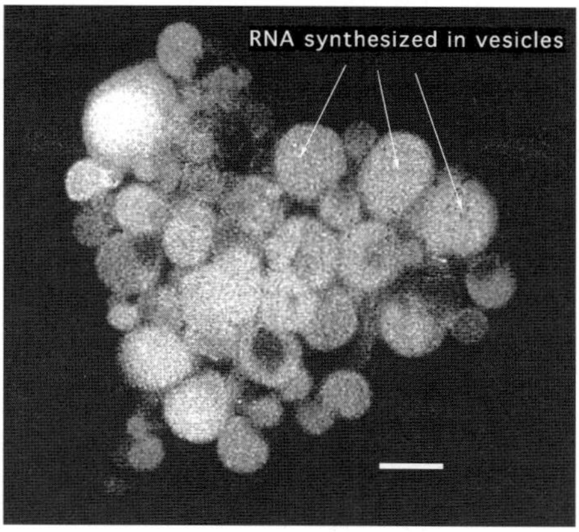

Fig. 11 Lipid vesicles with encapsulated T7 RNA polymerase and DNA template. A mixture of four nucleoside triphosphates was added, and these diffused into the vesicles and were used by the polymerase to synthesize RNA with DNA as a template. The RNA was stained with ethidium bromide and appears as fluorescent material within the vesicles. Note that some of the vesicles do not contain fluorescent RNA, presumably because they lacked sufficient enzyme or template. *Scale bar* shows 20 μm

substantially by using larger vesicles, demonstrating that green fluorescent protein can be synthesized by an encapsulated translation system. Noireaux and Libchaber [87] took this approach one step further by incorporating two genes into a similar encapsulated translation system, one for alpha hemolysin, a pore-forming protein, and a second for GFP. The investigators reasoned that if the system was in fact capable of synthesizing proteins, the newly translated hemolysin would insert into the membrane and produce a pore. This would allow externally added solutes (i.e., amino acids) to permeate the lipid bilayer barrier and supply the substrates required for protein synthesis. The production of GFP would then indicate that synthesis was proceeding at appreciable rates. This worked very well, and GFP could be visualized accumulating in the vesicular volume for up to 4 d.

The next obvious step is to incorporate both a gene transcription system and protein synthesis in lipid vesicles. This was reported in 2004 by Ishikawa et al. [87], who managed to assemble a two-stage genetic network in liposomes in which the gene for an RNA polymerase was expressed first and the polymerase then used to produce mRNA required for GFP synthesis.

The final challenge in modeling such systems will be to encapsulate an evolving ribozyme system [74, 86, 87] within vesicles formed from amphiphilic mixtures that are optimized for stability and permeability. It seems likely that one such mixture will have a set of properties that permit it to encapsulate a catalytic polymerase system and template, with sufficient permeability to allow substrate access to the enzyme at reasonable rates. Replication and ribozyme evolution would then occur in immensely large numbers of microscopic volumes represented by the liposome interiors, rather than in the macroscopic volume of a test tube. Under these conditions, the rare ribozyme that happens to undergo a favorable mutation would be readily selected, whereas in a test tube it is lost among trillions of other similar molecules.

13 Summary

This chapter provides a perspective on the most primitive forms of cellular life. In the early Earth environment, there must have been a variety of amphiphilic hydrocarbon derivatives that could self-assemble into bilayer boundary structures and encapsulate polymers that were being synthesized by a separate process. The vesicle membranes would have been sufficiently permeable to allow passage of smaller ionic substrates required for metabolism and biosynthesis, yet maintain larger molecules within. Encapsulated catalysts and information-bearing molecules would thus have had access to nutrients required for growth. Furthermore, specific groupings of macromolecules would be maintained, rather than drifting apart. This would allow true Darwinian-type selection of such groupings to occur, a process that could not take place in mixtures of molecules free in solution. A small number of the encapsulated molecular systems were likely to have the specific set of properties that allowed them to capture free energy and nutrients from their environment and undergo growth by polymerization. At some point, the growth would become catalyzed by the encapsulated polymers and then begin to be directed by a primitive genetic process. Such structures would be on the evolutionary path to the first forms of cellular life.

Acknowledgements We dedicate this chapter to the memory of John Oró, an early champion of the exogenous delivery of organic material and the central role of membranes in the origin of life. We wish to acknowledge the NASA Exobiology program (DD) and the NASA Astrobiology Institute (JD) for financial support.

References

1. Dyson F (1999) The Origins of Life. Princeton University Press, Princeton, NJ
2. Deamer D, Dworkin JP, Sandford SA, Bernstein MP, Allamandola LJ (2002) Astrobiology 2:371–382
3. Segré S, Deamer DW, Lancet D (2001) Orig Life Evol Biosphere 31:119–145
4. Cavalier-Smith T (1987) Cold Spring Harbor Symposia on Quantitative Biology, Vol LII, pp 805–824
5. Koch AL, Schmidt TM (1991) J Mol Evol 33:297–304

6. Morowitz HJ (1992) Beginnings of Cellular Life. Yale University Press, New Haven, CT
7. Miller SL (1953) Science 117:528–529
8. Miller SL, Urey HC (1959) Science 130:245–251
9. Miller SL, Schlesinger G (1984) Orig Life 14:83
10. Nooner DW, Gilbert JM, Gelpi E, Oró J (1976) Geochim Cosmochim Acta 40:915–24
11. McCollom TM, Ritter G, Simoneit BRT (1999) Orig Life Evol Biosphere 29:153–166
12. Rushdi AI, Simoneit B (2001) Orig Life Evol Biosphere 31:103–118
13. Kvenvolden KA, Lawless JG, Pering K, Peterson E, Flores J, Ponnamperuma C, Kaplan IR, Moore C (1970) Nature 28:923
14. Cronin JR, Pizzarello S, Cruikshank DP (1988) In: Kerridge JF, Matthews MS (eds) Meteorites and the Early Solar System. University of Arizona Press, Tucson, AZ, p 819–857
15. Sephton MA (2002) Nat Prod Rep 19:292–311
16. Holland HD (1984) The Chemical Evolution of the Atmosphere and Oceans. Princeton University Press, Princeton, NJ
17. Kasting JF, Brown LL (1998) In: Brack A (ed) The Molecular Origins of Life. Cambridge University Press, Cambridge, UK, pp 35–56
18. Oró J (1961) Nature 190:389–390
19. Delsemme A (1984) Orig Life 14:51–60
20. Anders E (1989) Nature 342:255–257
21. Chyba CF, Sagan C (1992) Nature 355:125–13
22. Maurette M (1998) In: Brack A (ed) The Molecular Origins of Life. Cambridge University Press, Cambridge, UK, pp 147–186
23. Ehrenfreund P, Charnley SB (2000) Ann Rev Astron Astrophys 38:427–483
24. Sandford SA (1996) Meteoritics Planet Sci 31:449–476
25. Greenberg M, Mendoza-Gomez CX (1993) In: Greenberg M, Mendoza-Gomez CX, Pironella V (eds) The Chemistry of Life's Origins. Kluwer, Dordrecht, pp 1–32
26. Bernstein MP, Sandford SA, Allamandola, LJ, Chang S, Scharberg MA (1995) Astrophys J 454:327–344
27. Ehrenfreund P, d'Hendecourt L, Charnley SB, Ruiterkamp R (2001) J Geophys Res 106:33291–33302
28. Muñoz-Caro GM, Meierhenrich WA, Schutte WA, Barbier B, Arcones Segovia A, Rosenbauer W, Thriemann HP, Brack A, Greenberg JM (2002) Nature 416:403–406
29. Bernstein MP, Dworkin JP, Sandford SA, Cooper GW, Allamandola LJ (2002) Nature 416:401
30. Bernstein MP, Dworkin JP, Sandford SA, Allamandola LJ (2001) Meteoritics Planet Sci 36:351–258
31. Dworkin JP, Deamer DW, Sandford SA, Allamandola LJ (2001) Proc Natl Acad Sci USA 98:815–819
32. Pierazzo E, Chyba C (1999) Meteoritics Planet Sci 32:090–918
33. Krishnamurthy RV, Epstein S, Cronin JR, Pizzarello S, Yuen GU (1992) Geochim Cosmochim Acta 56:4045–4058
34. Sandford SA, Bernstein MP, Dworkin JP (2001) Meteoritics Planet Sci 36:1117–1133
35. Love SG, Brownlee, DE (1993) Science 262:550–553
36. Hargreaves WW, Mulvihill SJ, Deamer DW (1977) Nature 266:78–80
37. Rao M, Eichberg MR, Oró J (1982) J Mol Evol 18:196–202
38. Epps DE, Sherwood E, Eichberg J, Oró J (1978) J Mol Evol 6:279–92
39. Ourisson G, Nakatani T (1994) Chem Biol 1:11
40. Conde-Frieboes K, Blochliger E (2001) Biosystems 1:109–114
41. Singer SJ, Nicolson GL (1972) Science 175:720–31

42. Frye LD, Edidin M (1970) J Cell Sci 7:319–35
43. Oliver A, Deamer DW (1994) Biophys J 66:1364–79
44. Vlassov A, Khvorova A, Yarus M (2001) Proc Natl Acad Sci USA 98:7706
45. Pohorille A, Schweighofer K, Wilson MA (2005) Astrobiology 1–17
46. Chakrabarti A, Deamer DW (1994) J Mol Evol 39:1–5
47. Parsegian A (1969) Nature 221:844–846
48. Paula S, Volkov AG, Van Hoek AN, Haines TH, Deamer DW (1996) Biophys J 70:339–348
49. Monnard P-A, Deamer DW (2001) Orig Life Evol Biosphere 31:147–155
50. Walde P, Wick R, Fresta M, Mangone A, Luisi PL (1994) J Am Chem Soc 116:11649–11654
51. Hanczyc MM, Fujikawa SM, Szostak JW (2003) Science 302:618–22
52. Hanczyc MM, Szostak JW (2004) Curr Opin Chem Biol 28:660–664
53. Shew R, Deamer D (1983) Biochim Biophys Acta 816:1–8
54. Pick U (1981) Arch Biochem Biophys 212:186
55. Nasseau M, Boublik Y, Meier W, Winterhalter M, Fournier D (2001) Biotech Bioeng 75:615
56. Stribling R, Miller SL (1987) Orig Life Evol Biosphere 17:261–73
57. Cairns-Smith G (1982) Genetic Takeover and the Mineral Origins of Life. Cambridge University Press, Cambridge, UK
58. Hazen RM, Filley TR, Goodfriend GM (2001) Proc Natl Acad Sci USA 98:5487–5490
59. Corliss JB, Baross JA, Hoffman SE (1981) Oceanol Acta. Proceedings of the 26th International Geological Congress, Paris, pp 59–69
60. Baross JA, Hoffman SE (1985) Orig Life 15:327
61. Pace NR (1991) Cell 65:531–533
62. Wächtershäuser G (1988) Syst Appl Microbiol 10:207–210
63. Wächtershäuser G (1988) Microbiol Rev 52:452–484
64. Huber C, Wächtershäuser G (1997) Science 276:245
65. Huber C, Wächtershäuser G (1998) Science 281:670–672
66. Martin W, Russell MJ (2003) Phil Trans R Soc Lond B 358:59–83
67. Lawless JG, Yuen GU (1979) Nature 282:396–398
68. Naraoka H, Shimoyama A, Komiya M, Harada H (1999) Orig Life Evol Biosphere 29:187–201
69. Deamer DW (1985) Nature 317:792–794
70. Deamer DW, Pashley RM (1989) Orig Life Evol Biosphere 19:21–33
71. Hargreaves WR, Deamer DW (1978) Biochemistry 17:3759–3768
72. Apel CL, Deamer DW, Mautner M (2002) Biochim Biophys Acta 1559:1
73. Monnard P-A, Apel CL, Kanavarioti A, Deamer DW (2002) Astrobiology 2:139
74. Johnston WK, Unrau PJ, Lawrence MS, Glasner ME, Bartel DL (2001) Science 292:1319–1325
75. Luisi PL (1996) Adv Chem Phys 92:425–438
76. Pohorille A, Deamer DW (2002) Trends Biotechnol 20:123
77. Szostak JW, Bartel DP, Luisi PL (2001) Nature 409:387–390
78. Rasmussen S, Chen L, Deamer D, Krakauer DC, Packard NH, Stadelr PF, Bedau MA (2004) Science 303:963–5
79. Chakrabarti A, Breaker RR, Joyce GF, Deamer DW (1994) J Mol Evol 39:555–559
80. Walde P, Goto A, Monnard P-A, Wessicken M, Luisi PL (1994) J Am Chem Soc 116:7541–7547

81. Oberholzer T, Wick R, Luisi PL, Biebricker CK (1995) Biochem Biophys Res Commun 207:250
82. Oberholzer T, Albrizio M, Luisi PL (1995) Curr Biol 2:677
83. Monnard P-A, Deamer DW (2002) Anat Rec 268:196
84. Yu W, Sato K, Wakabayashi M, Nakaishi T, K-Mitamura EP, Shima Y, Urabe I, Yomo T (2001) J Biosci Bioengin 92:590
85. Nomura S, Tsumoto K, Hamada T, Akiyoshi K, Nakatani Y, Yoshikawa K (2003) Chem Biochem 4:1172-1175
86. Noireaux V, Libchaber A (2004) Proc Natl Acad Sci USA 101:17669–74
87. Ishikawa K, Sato K, Shima Y, Urabe I, Yomo T (2004) FEBS Lett 576:387
88. Beaudry AA, Joyce GF (1992) Science 342:255
89. Wilson C, Szostak JW (1994) Nature 374:777–782

Top Curr Chem (2005) 259: 29–68
DOI 10.1007/b136152

Published online: 10 August 2005

On the Prebiotic Synthesis of Nucleobases, Nucleotides, Oligonucleotides, Pre-RNA and Pre-DNA Molecules

Raffaele Saladino[1] (✉) · Claudia Crestini[2] · Giovanna Costanzo[3] · Ernesto DiMauro[4]

[1]INFM, Dipartimento A.B.A.C., Università della Tuscia, 01100 Viterbo, Italy
saladino@unitus.it

[2]Dipartimento di Scienze e Tecnologie Chimiche, Università "Tor Vergata", 00133 Rom, Italy

[3]Istituto di Biologia e Patologia Molecolari, CNR, 00185 Rom, Italy

[4]Fondazione "Istituto Pasteur", Fondazione Cenci-Bolognetti, Università "La Sapienza", 00185 Rom, Italy

1 **Introduction** 30

2 **Prebiotic Synthesis of Purine and Pyrimidine Nucleobases** 32
2.1 Purine Nucleobases 32
2.2 Pyrimidine Nucleobases 37
2.3 Prebiotic Synthesis of Purine and Pyrimidine Nucleobases Under Catalytic Conditions 40

3 **Prebiotic Synthesis of Purine and Pyrimidine Nucleosides** 44

4 **Prebiotic Synthesis of Purine and Pyrimidine Nucleotides** 47

5 **Prebiotic Synthesis of Oligonucleotides** 52

6 **Pre-RNA and Pre-DNA Molecules** 56

7 **Concluding Remarks** 61

References 62

Abstract All the strategies for the prebiotic syntheses of RNA and DNA assume the adequate availability of the presumptive precursors such as purine and pyrimidine nucleic bases, nucleosides and nucleotides. Polymerization of activated nucleotides probably furnished the first informational oligonucleotides. The formation of these precursors from mixtures of simple gases was shown to occur in a variety of conditions including UV-irradiation, electric discharge, heating, volcanic activity and marine vents. Even though a well-defined physical chemical scenario on the primitive Earth is not available, prebiotic syntheses were most probably performed using the simplest chemicals and the most common conditions present at that time. During these synthetic processes minerals played the relevant role of catalysts furnishing local microenvironments for the selective concentration of reagents and for the preservation of newly formed products. Here we focus on the optimal experimental conditions needed to carry out these syntheses and on the characterization of the major products thus obtained. Special attention will be addressed to catalyzed processes. Taken together, these procedures and data suggest novel scenarios for the molecular evolution of life on the primitive Earth and may provide a chemical clue to the evaluation of the plausible emergence of extraterrestrial forms of life.

Keywords Origin of life · Prebiotic synthesis · Nucleic bases · Nucleotides · Oligonucleotides

Abbreviations

AMN	Aminomaleonitrile
DAMN	Diaminomaleonitrile
AICN	4-Aminoimidazole-5-carbonitrile
AICA	4-Aminoimidazole-5-carboxamide
5′-IMP	Inosine-5′-monophosphate
Al – PILC	Aluminum pillared clays
2′-AMP	Adenosine-2′-monophosphate
3′-AMP	Adenosine-3′-monophosphate
2′,3′-cAMP	Adenosine 2′,3′-cyclic phosphate
5′-ATP	5′-Adenosinetriphosphate
DISN	Diiminosuccinonitrile
5′-GMP	Guanosine-5′-monophosphate
ImpC	5′-Phosphorimidazolide cytidine
D-2-MeImpG	D-guanosine 5′-phosphoro-2-methylimidazole
L-2-MeImpG	L-guanosine 5′-phosphoro-2-methylimidazole
APH	Alkaline phosphatase
Rnase T_2	Ribonuclease T_2
TNA	L-α-threofuranosyl-(3′-2′)-oligonucleotides
(homo-DNA)	2′,3′-Dideoxy-glucopyranosyl-(6′-4′)-β-oligonucleotides
HNAs	Hexitol nucleic acids
ANAs	Altritol nucleic acids
PNA	Peptide nucleic acid

1 Introduction

The prebiotic chemistry deals with the emergence of the first biomolecules considered in their possible function of self-assembling building blocks, eventually leading to the last common ancestor on the primitive Earth [1, 2]. From a geological point of view the evolution from organic compounds to primitive organisms probably occurred in the relatively short time of 0.4×10^9 years [3]. In the specific context of the prebiotic chemistry of nucleic acids, major efforts have been devoted to the definition of the chemical structures, reactions and synthetic pathways of the first molecules carrying out the functions that had to become the present-day DNA and RNA-based genetic apparatus. Considering the known structures of current nucleic acids, the main topics to be analyzed in this context are the synthesis of purine and pyrimidine nucleobases, the formation of the glycosidic bonds between bases and sugars and their polymerization to oligonucleotides by activated phosphates. Because of the high pharmacological activity of nucleic acid derivatives, mainly as antiviral and antitumoral drugs, thousands of different syntheses of natural and unnatural purines and pyrimidines, nucleosides and nucleotides have been described. Thus, in order to establish the prebiotic role of these syntheses, stringent criteria are necessary. To date, detailed plausible chemical scenarios for the primitive Earth are not available [4, 5]. Operationally, we propose to consider as prebiotic any synthetic chemical event that occurred starting from simple precursors (one-carbon or two-carbon atoms containing fragments) under gaseous, solution or solid-state conditions. In this respect, one-pot transformations are obviously the more interesting. The simplest precursors for nucleic acids were most probably components of the primordial atmosphere [6], even though products of hydrothermal sources, "black smokers" activities [7–9], or compounds coming from the interstellar space by meteoritic and cometary depositions cannot be ruled out [10–12]. In this setting, the energy necessary for the condensation reactions of precursors to more complex derivatives was presumably provided by high potential electric discharges, ultraviolet radiation, redox transformations, heat from hydrothermal and volcanic activities [13–15]. Since the hypothesis of a "primitive soup" of precursors suggested by Oparin [16], different one-carbon atom fragments, such as hydrogen cyanide (HCN), cyanate (NCO^-), cyanogens ($(CN)_2$), formaldehyde (HCOH), formamide (NH_2CHO), formic acid (HCOOH), ammonium formate ($NH_4^+HCOO^-$), ammonium cyanide (NH_4CN), urea (NH_2CONH_2) and two- or three-carbon atoms containing fragments, such as cyanoacetylene, cyanoacetaldehyde and β-alanine, were selected as starting material for the synthesis of nucleic acid components and sugars [17–20]. These compounds can be easily obtained by spark discharge or shock waves on gaseous mixtures of methane (CH_4), carbon oxides (CO and CO_2), ammonia, nitrogen and water [21–28], and they can be interconverted by hydrolytic processes [29, 30]. In addition, several one-carbon containing precursors are ubiquitous molecules detected in comets, in interplanetary dust, in the atmosphere of planets, and on the surface of moons [31–36]. Noteworthy, the prebiotic precursors of nucleic acids and their components are also involved in the synthesis of other relevant biomolecules such as amino acids. Several amino acids were for instance obtained from aldehydes, hydrogen cyanide and ammonia by a Strecker mechanism [37] during the spark discharge experiments on simple gaseous mixtures carried out by Miller in the early 1950s [38, 39]. We describe here the different experimental approaches for the prebiotic synthesis of purine and pyrimidine nucleobases, nucleosides, nucleotides, oligonucleotides and pre-RNA and pre-DNA molecules, starting from the most efficient and widely diffused chemical precursors. We will in particular focus on the optimal experimental conditions needed to carry out these syntheses and on the characterization of the major products thus obtained. Special attention will be addressed to catalyzed processes. Taken together, these procedures and data suggest novel scenarios for the molecular evolution of life on the primitive Earth and may provide a chemical clue to the evaluation of the plausible emergence of extraterrestrial forms of life.

2 Prebiotic Synthesis of Purine and Pyrimidine Nucleobases

2.1 Purine Nucleobases

In the early 1960s, Orò described for the first time the one-pot synthesis of adenine **1** from a concentrated mixture of HCN (1.0–11 M solutions of HCN) and ammonia [40–42].

All the carbon atoms of the purine ring were supposedly provided by HCN molecules through a complex step-by-step condensation process. In particular, oligomers of HCN, such as the HCN-trimer aminomaleonitrile (AMN) and the HCN-tetramer diaminomaleonitrile (DAMN), were found to be intermediates in this transformation (Scheme 1) [43, 44]. In accordance with the present-day biosynthesis of purines in the cell, two 4,5-disubstituted imidazole derivatives, 4-aminoimidazole-5-carbonitrile (AICN) and 4-aminoimidazole-5-carboxamide (AICA) were successively formed from AMN and DAMN by chemical or, most probably, photochemical reactions [45–47]. Finally, a ring-closure process of AICA and HCN yielded adenine **1**.

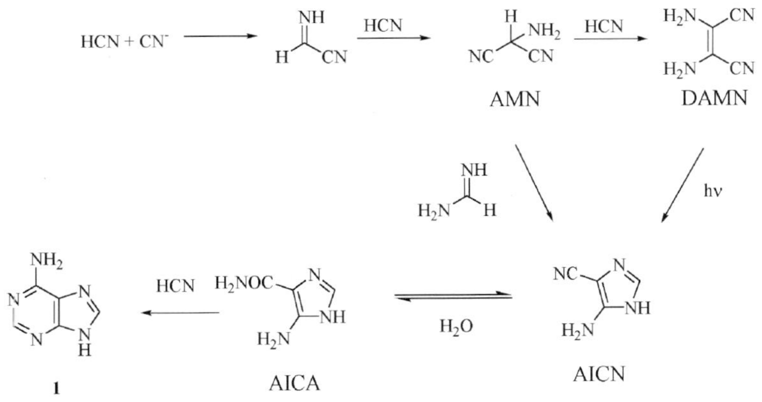

Scheme 1

This chemical procedure was found to be of general applicability and a large panel of purine derivatives, namely hypoxanthine **2**, diaminopurine **3**, xanthine **4**, isoguanine **5** and guanine **6** were obtained from the imidazole intermediates in the presence of other one-carbon fragments such as cyanide and cyanogens (Scheme 2).

In agreement with the "chemomimetic" concept as defined by Eschenmoser, the panel of enzymatic transformations for the biosynthesis of purines that we currently observe in the cell can be hypothesized to have evolved from primitive chemical processes [48–50]. 2-Carbonitrile and 2-carboxamide AICA and AICN derivatives, respectively, were also used as intermediates for the synthesis of adenine **1** and 8-substituted adenines **7** and **8** [51]. In principle, purine derivatives **7** and **8** may pair with pyrimidine bases by formation of Watson–Crick or Hoogsteen hydrogen bond interactions.

A HCN-pentamer, formed by reaction of *N*-(methylimino)acetonitrile and AMN was also suggested as precursor of adenine **1** (Scheme 3).

Since high concentrations of HCN were not plausibly present on the primitive Earth, several experiments were performed using dilute solutions of HCN in alkaline media at high temperature (< 0.1 M HCN solutions) [52]. Under these experimental conditions, the purine and pyrimidine derivatives were not directly recovered from the reaction mixtures, the chemical information

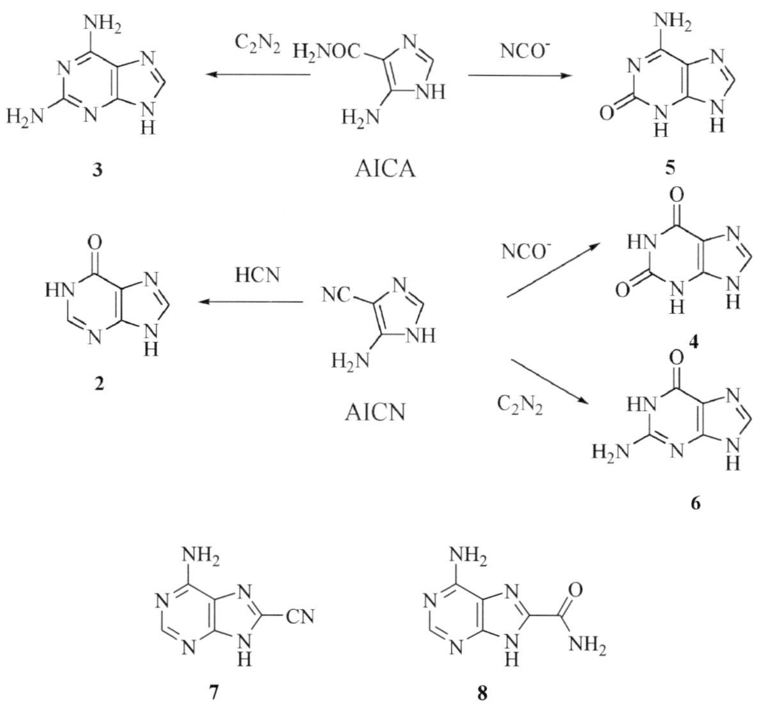

Scheme 2

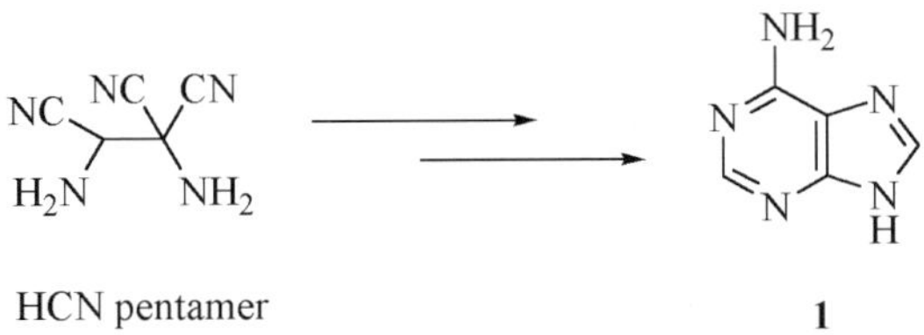

Scheme 3

necessary for their synthesis being stored inside stable and low-solubility HCN-polymers named "alzulmins" [53–55].

On the basis of several analytical studies (differential thermal analysis, fluorescence, CPMAS solid-state NMR spectroscopy and others) [56–58] two models have been proposed to describe the structure of HCN-polymers, the Umemoto [59] and the Völker models [60]. In the Völker model, HCN polymerizes to extensive double-ladder rod-like structures, while a simpler mono-ladder pattern was hypothesized by Umemoto (Fig. 1). Irrespective of the structure assumed by HCN-polymers, a large panel of purine, imidazole and pyrimidine derivatives can be obtained by hydrolysis of these materials. In 1963, Lowe described the first example of acidic hydrolysis of the HCN-polymer (boiling 6.0 N HCl) to yield amino acids, carboxylic acids, adenine and hypoxanthine (Scheme 4).

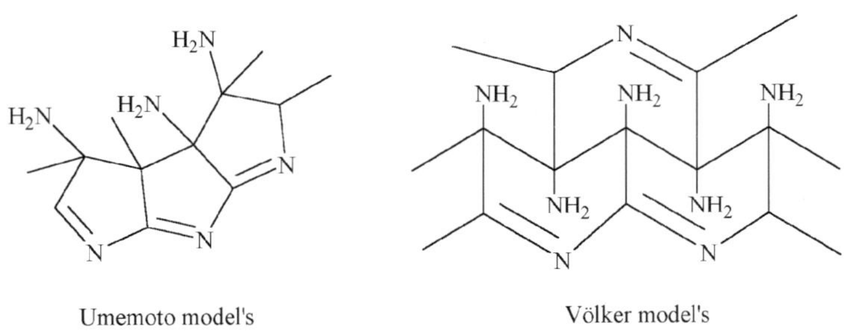

Fig. 1 Schematic representations of the Umemoto and Völker models for HCN-polymers

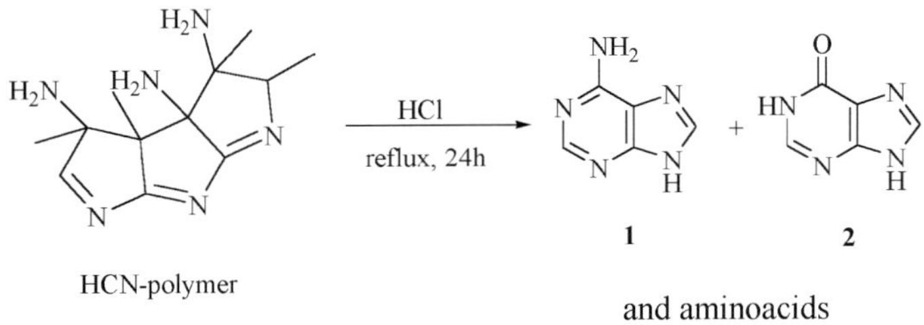

Scheme 4

Successively, Ferris and co-workers performed the hydrolysis of HCN-polymer both under acidic and alkaline conditions [61].

Treatment of HCN-polymer with 6.0 N HCl afforded adenine **1**, AICN, 3,4-dihydroxypyrimidine **9** and 5-hydroxyuracil **10**, while **1**, AICN and orotic acid **11** were recovered after reaction with sodium hydroxide (Scheme 5). A reaction mechanism involving the formation of different aminopyridines as intermediates and reduction steps was proposed to explain the distribution of the obtained products. In accordance with the "chemomimetic" concept, orotic acid is a key intermediate in the current biosynthesis of pyrimidine nucleotides [62].

Remarkably, HCN-polymers are efficiently obtained at low temperatures suggesting a plausible cold origin-of-life scenario [63–65]. This latter hypothesis provides a mechanism to account for the concentration of HCN and a more stable environment for the newly formed nucleic acid components. Moreover, HCN is easily concentrated at its eutectic temperature (– 23.4 °C) [66, 67] and the production of DAMN was found to be accelerated by lowering the temperature of the reaction medium. A large panel of purine derivatives, adenine **1**, hypoxanthine **2**, diaminopurine **3**, xanthine **4** and guanine **5** was obtained after acidic and alkaline hydrolysis at high temperatures of the HCN-polymer produced in a frozen ammonium cyanide solution at – 78 °C for 27 years (Scheme 6; yields of products are reported

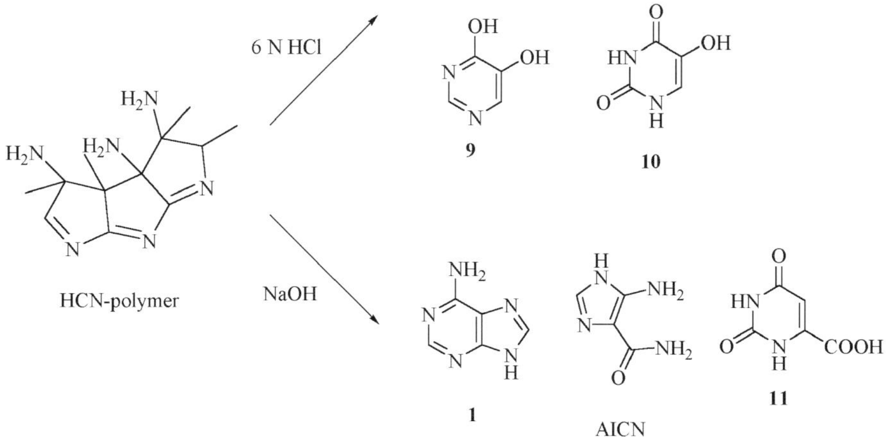

Scheme 5

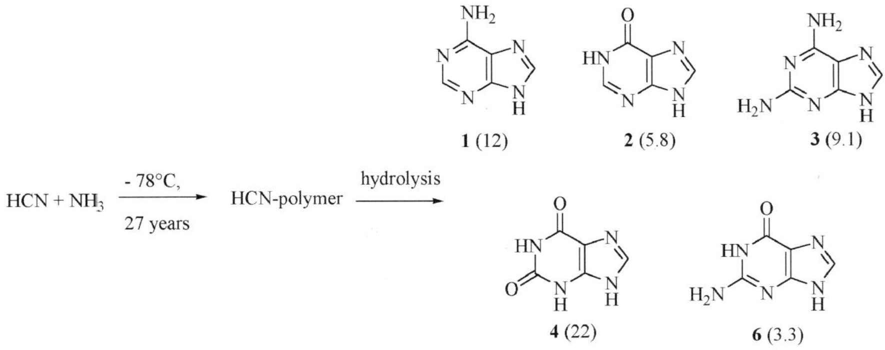

Scheme 6

in parenthesis) [68, 69]. Different pyrimidine derivatives were also detected in the reaction mixture, 3,4-dihydroxypyrimidine being the most abundant product. In the absence of the hydrolysis step only a low yield of products was obtained. A structural comparison between the HCN-polymers formed at high or low temperature is not available.

Purine derivatives were also synthesized starting from HCN derivatives. Yamada described in 1972 the one-pot synthesis of purine **12** obtained simply by heating neat formamide, a product of HCN hydrolysis [70, 71] at 160 °C [72]. Adenine **1** was obtained as the main reaction product along with a low amount of **12** upon repetition of the reaction under similar experimental conditions in the presence of HCN (Scheme 7) [73, 74]. On the basis of $^{13}C-NMR$ studies of the synthesis performed with the enriched substrate, three molecules of HCN and two molecules of formamide were found to be incorporated into the adenine scaffold by a C – N bond fission process.

The formation of purine from ammonium formiate ($NH_4^+HCOO^-$), again a product of hydrolysis of HCN, was reported by Zubay and co-workers [75]. Treatment of AICN and AICA with a concentrated solution of $NH_4^+HCOO^-$ at 93 °C afforded adenine **1** and hypoxanthine **2**, respectively (Scheme 8). A reaction pathway in which formate and ammonium ions were both reactants has been hypothesized for the formation of adenine.

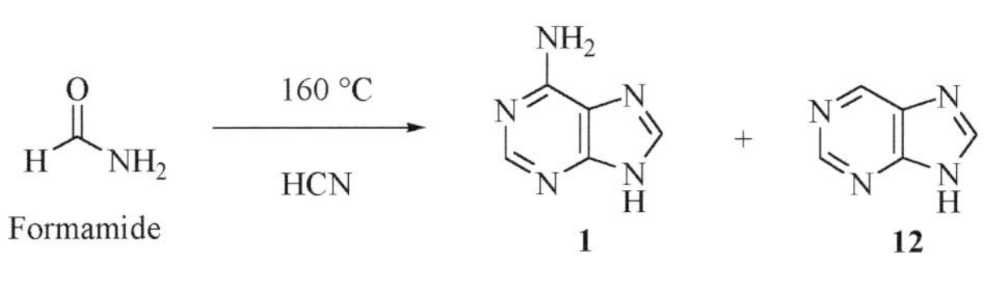

Scheme 7

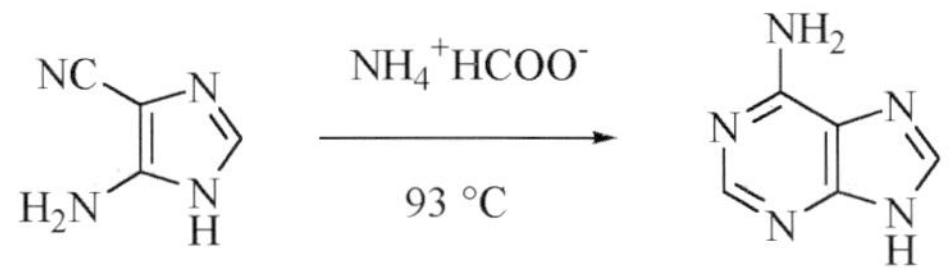

AICN 1

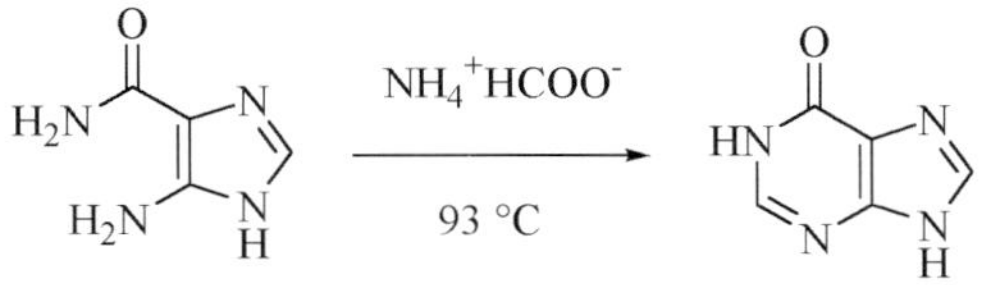

AICA 2

Scheme 8

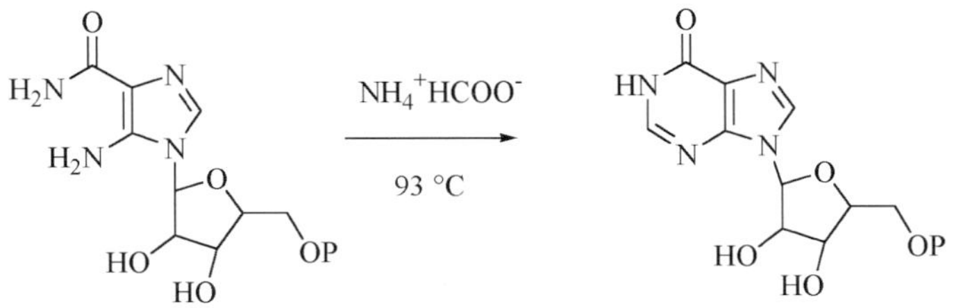

13 14

P= Phosphate

Scheme 9

Similar results were obtained in the conversion of 5′-phosphoribosyl-aminoimidazole carboxamide **13** to inosine-5′-monophosphate **14** (5′-IMP, Scheme 9).

Ammonium formate was also used in the synthesis of adenine **1** from DAMN [76]. Irrespective of the experimental conditions used for the synthesis of purine nucleobases, only a few procedures for the preparation of guanine have been reported.

2.2 Pyrimidine Nucleobases

In spite of the large amount of pyrimidines found in the Murchison meteorite (one fifth of that of purines), only a few instances of prebiotic synthesis of these derivatives have been reported, with the exception of the catalyzed procedures that will be described in the next paragraph. In general, these procedures are characterized by low yields. In this context, two main prebiotic precursors have been identified, cyanoacetylene **15** and cyanoacetaldehyde **16**. These compounds contain a preformed C – C bond which is incorporated in the C5-C6 position of the pyrimidine ring. In 1968 Ferris and co-workers reported that the reaction of **15** with cyanate (NCO^-) at 30 °C yields cytosine **17** and, after its hydrolysis, uracil **18** in acceptable yield (Scheme 10). Cytosine is readily hydrolyzed to uracil showing a $t_{1/2} = 200$ years at 30 °C in neutral solution [77]. *Trans*-cyanovinylurea **16** was recovered as a key intermediate for this transformation [78].

However, this reaction requires relatively high concentrations of cyanate (> 0.1 M), unlikely to occur in aqueous media given its rapid degradation to carbon dioxide and ammonia. In order to optimize the yield of cytosine by this procedure, cyanoacetylene **15** was treated with a lower amount of cyanate at 100 °C. Under these more efficient experimental conditions cytosine was obtained in 19% overall yield along with a low amount of uracil [79]. An alternative route explored consists of the use of urea **19**, which is considered a common prebiotic precursor [80], as one-carbon atom donor in the place of cyanate. The reaction of cyanoacetylene **15** (1.0 M) with **19** (1.0 M) at 100 °C gave cytosine **17** in 5% yield (Scheme 11). At low concentrations of **19** (< 0.1 M) this reaction does not produce detectable amounts of cytosine [81].

A similar pattern of reactions was studied starting from cyanoacetaldehyde **20**, the first product to be formed during the hydrolysis of cyanoacety-

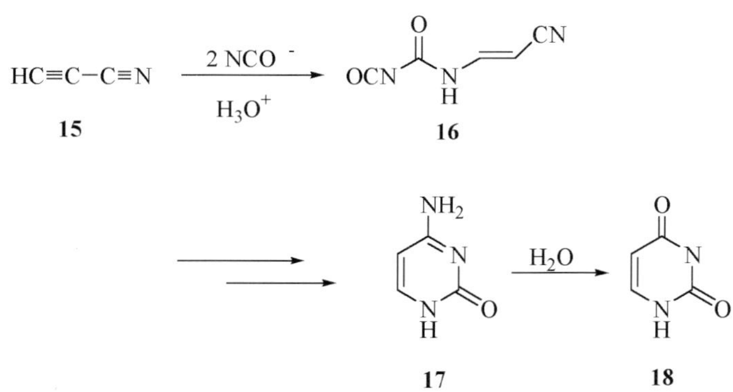

Scheme 10

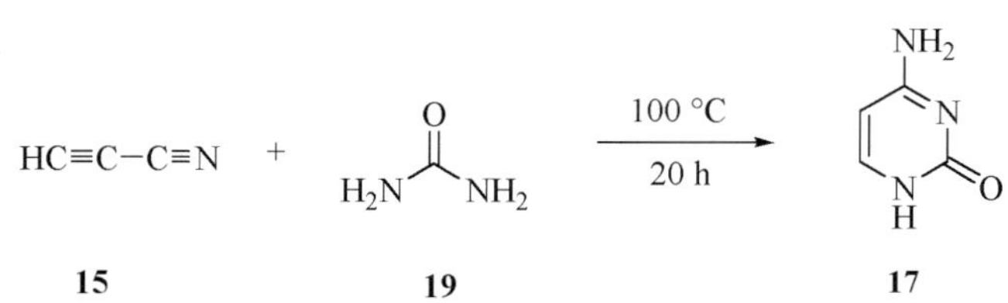

Scheme 11

lene. Cyanoacetaldehyde can be also formed by ultraviolet irradiation of aqueous formaldehyde [82]. Cyanoacetaldehyde **20** reacts with concentrated solutions of urea **19** at 100 °C to give cytosine **17** in 53% yield (Scheme 12) [83]. A drop in cytosine yield was observed for prolonged reaction times probably due to its hydrolysis to uracil. Under similar experimental conditions and with thiourea **21** as condensing agent 2-thiocytosine **22**, one of the minor components of present-day tRNA was obtained (Scheme 12).

A related synthesis was performed using guanidine **23** in the place of urea to give 2,4-diaminopyridine **24** in low yield (0.1–2.0%) [84]. Cytosine was successively obtained from **24** by hydrolysis (Scheme 13).

In principle, this synthesis is preferred over that from cyanoacetylene and cyanate because of the great stability of the reactants (the stability of cyanoacetaldehyde is 103 times that of cyanoacetylene and the half-life for guanidine hydrolysis is 105–108 years at pH 9.0). More recently, better results (40–85% yield of cytosine) were obtained performing the reaction with a concentrated solution of reagents mimicking the drying lagoon model of prebiotic synthesis [85]. Noteworthy, pyrimidine nucleobases have also been obtained starting from amino acid derivatives. In the 1970s, Schwartz described the synthesis of uracil by ultraviolet irradiation of a mixture of cyanate and β-alanine **25** (Scheme 14). 5,6-Dihydrouracil **26**, a minor component of present-day tRNA, was recovered as intermediate of the transformation [86].

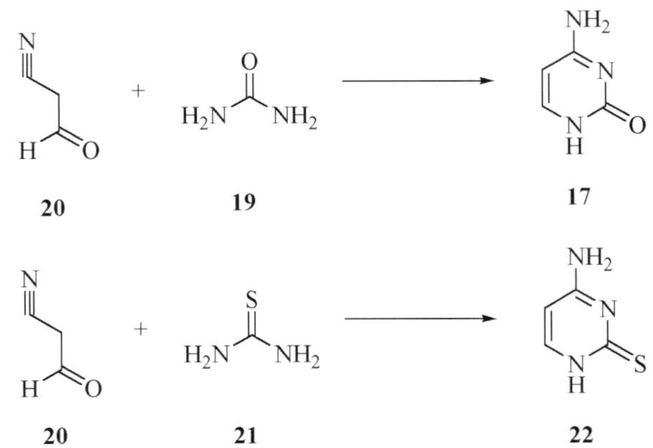

Scheme 12

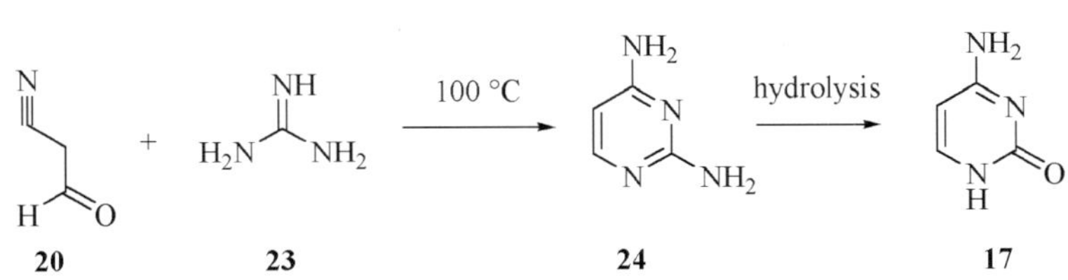

Scheme 13

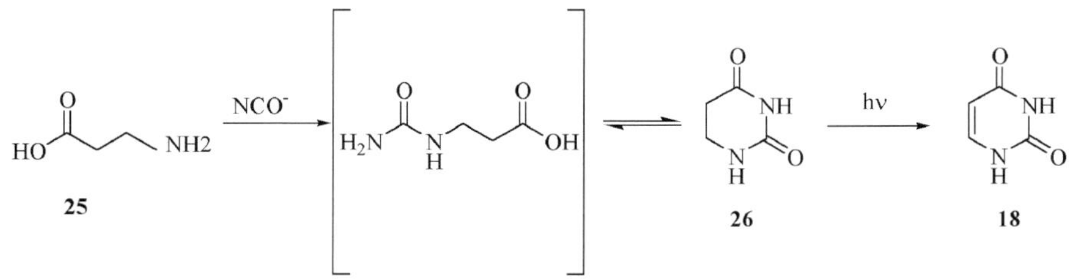

Scheme 14

2.3 Prebiotic Synthesis of Purine and Pyrimidine Nucleobases Under Catalytic Conditions

Catalysis probably played a crucial role in the prebiotic synthesis of purine and pyrimidine nucleobases. Several minerals and metal oxides characterized by chemical and photochemical catalytic properties were widely diffused on the primitive Earth. These compounds, when components of a reaction mixture, could have enhanced the efficacy of prebiotic syntheses affecting the selectivities of the reactions and providing local microenvironments able both to concentrate dilute reagents and to preserve newly formed nucleobases from degradative processes. Purine and pyrimidine nucleobases may, for instance, be differentially adsorbed on the surface of minerals such as pyrite (FeS_2), quartz (SiO_2), pyrrhotite (FeS), magnetite (Fe_3O_4), forsterite (Mg_2SiO_4) and graphite [87–89]. To date, the role of the catalysis on the prebiotic synthesis of nucleobases has been analyzed in some detail only in the case of formamide as prebiotic precursor. Formamide has a boiling point of 210 °C with limited azeotropic effects and, at difference from HCN, it can be easily concentrated by heating in lagoons and on drying beaches. Following the initial studies on the preparation of purine derivatives from neat formamide, the selectivity and the efficiency of the formamide condensation in the presence of catalysts were analyzed. In a first series of experiments, formamide was heated at 160 °C in the presence of catalytic amounts (10% in weight) of calcium carbonate, silica, alumina, zeolite (of Y type) or kaolin. Under these experimental conditions several purine and pyrimidine derivatives including adenine **1**, purine **12**, cytosine **17** and 4-hydroxypyrimidine [4(3H)pyrimidinone] **27** were recovered in different yields depending on the catalyst used for the transformation (Scheme 15) [90, 91].

Purine was the only recovered product in the presence of calcium carbonate. Zeolite and silica were the best catalysts for the synthesis of cytosine (c.a. 4.3 and 4.1 mg of cytosine per gram of formamide, respectively), while alumina gave the highest yield of 4-hydroxypyrimidine **27** (c.a. 2.0 mg of cytosine per gram of formamide).

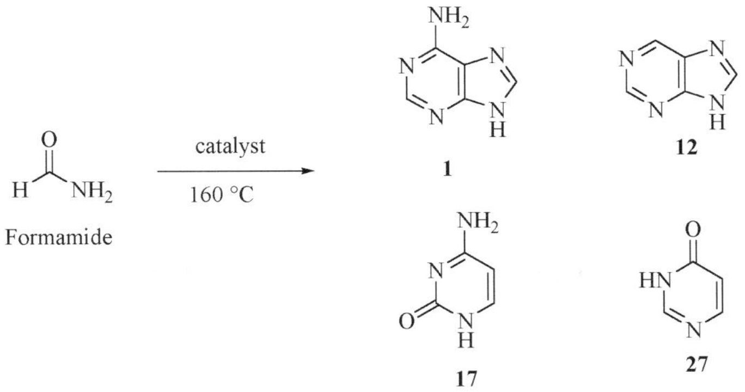

Scheme 15

Noteworthy, this is the first synthesis of cytosine reported starting from a one-carbon atom fragment as simple as formamide. Moreover, irrespective of the experimental conditions, cytosine and 4-hydroxypyrimidine were recovered in yield higher than adenine (major yield for adenine: c.a. 0.7 mg of adenine per gram of formamide; data obtained in the presence of silica). Probably, the HCN and ammonia necessary in addition to formamide for the formation of the adenine and cytosine scaffolds derive from a catalyzed partial degradation of formamide, usually occurring at higher temperatures [92]. Montmorillonites produced after acidic treatments also played an important role in the formamide condensation process. Montmorillonites are naturally occurring clays that show a high ability of swelling and exchanging ions [93]. Since Bernal's analysis of the physical basis of life [94], several prebiotic scenarios invoking clays have been proposed. Accordingly, these minerals have been used as microreactors to concentrate organic reagents, as templates for polymerizations [95] and protective environments for low-stability biomolecules [96]. When formamide was heated in the presence of montmorillonites K-10, K-30, KSF and aluminium-pillared clays (Al-PILC), which differ for pore size distribution, surface area, acidity and cation exchange capability, a large panel of nucleobases, adenine **1**, hypoxanthine **2**, purine **12**, cytosine **17**, uracil **18** and N^9-formylpurine **28** was obtained, in addition to nucleic acid precursors such as AICA and N-formylAICA (Scheme 16) [97]. Again, a different selectivity of the reaction was observed depending on the nature of the catalyst. Montmorillonite KSF was the best catalyst for cytosine (c.a. 18.0 mg of cytosine per gram of formamide), while the highest yield of adenine was obtained with montmorillonite K-30 (c.a. 22.0 mg of adenine per gram of formamide). N^9-Formylpurine **28**, which contains a masked glycosidic bond on its formyl moiety, was suggested to be a plausible precursor of purine acyclonucleosides. Being formamide able to degrade DNA [98], studies on the degradation of nucleobases under experimental conditions comparable to those of their synthesis

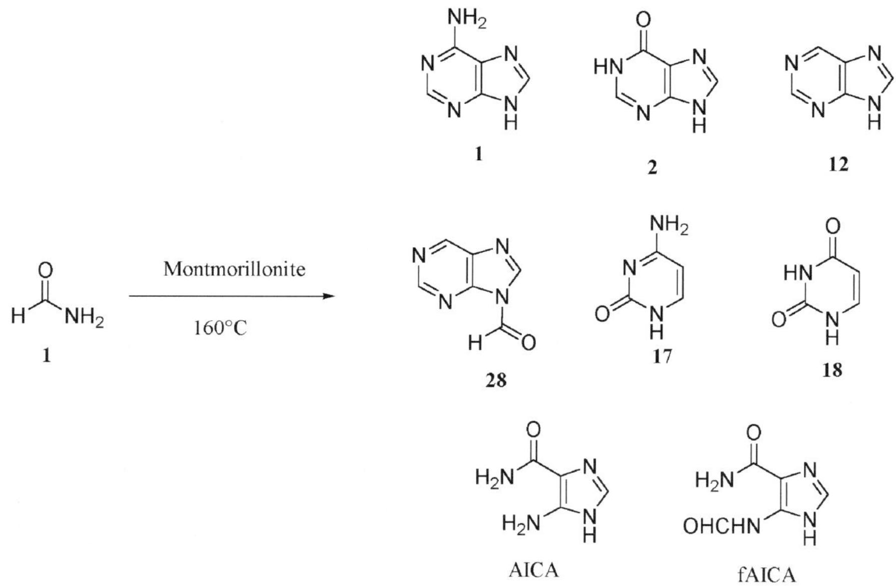

Scheme 16

were performed. Montmorillonites differentially affect the rate of degradation of nucleobases when embedded in 2′-deoxy-oligonucleotides. Adenine and guanine were protected from the described degradative action exerted by formamide [99], while the rate of degradation of thymine was unexpectedly enhanced. It is reasonable to suggest that in the case of the purine nucleobases the formation of complexes occurs between the N-7 position and/or the exoclycic NH_2 group of the purine ring and the metal ions present on the clay structure. In both cases the stereo-electronic properties of the purine bases relevant for the reaction with formamide are strongly modified, thus justifying a novel pattern of reactivity. In the case of pyrimidine nucleobases this effect was found to be less pronounced, probably because of the lower efficiency of pyrimidines in the complexation with clays.

Additional advantages of the formamide condensation protocol were observed performing the reaction in the presence of catalysts which decompose formamide to formaldehyde. Formaldeyde is the most important prebiotic precursor of sugars through a series of enolization and aldol-like condensation processes catalyzed under acidic or basic conditions, known as the "formose" reaction [100]. When formamide was heated at 160 °C in the presence of titanium dioxide (a catalyst able to degrade amides to aldehydes) [101] a complex mixture of nucleobase derivatives was obtained including adenine **1**, purine **12**, cytosine **17**, N^9-formylpurine **28**, N^9, N^6-diformyladenine **29**, 5-hydroxymethyluracil **30**, thymine **31** and three novel

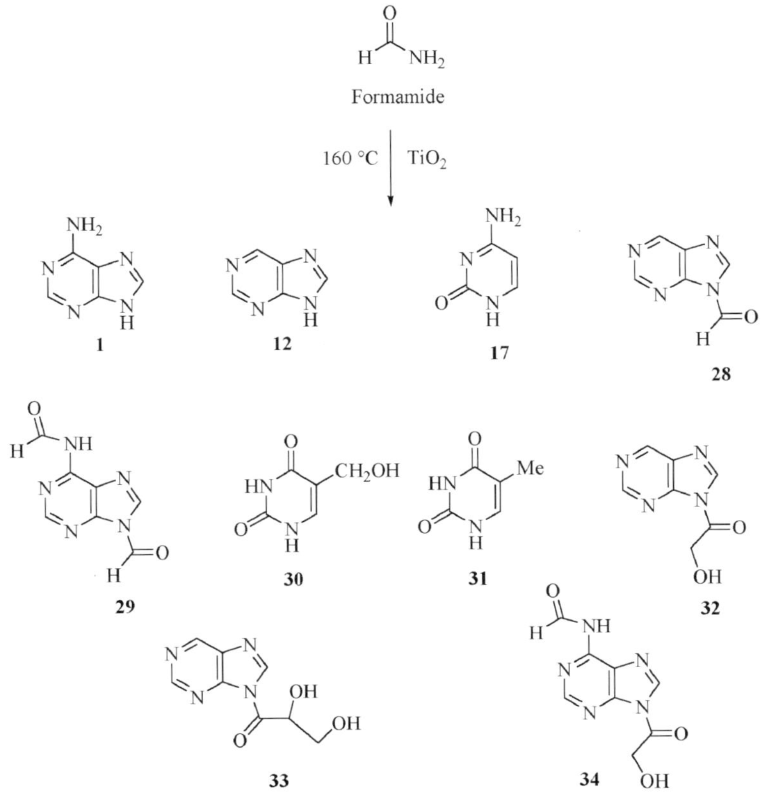

Scheme 17

purine and adenine acyclonucleosides, compounds **32–34** (Scheme 17). In the formation of 5-hydroxymethyluracil **30**, thymine **31** and purine acyclonucleosides **32–34**, which were never previously obtained from formamide, formaldehyde played an important role [102].

5-Hydroxymethyluracil **30**, a component of the present-day DNA of *Bacillus subtilis* bacteriophages [103], was obtained by electrophilic addition of formaldehyde to the C5-C6 double bond of a preformed uracil ring (which is probably the reason for the absence of uracil in the reaction mixture). Thymine was then obtained from 5-hydroxymethyluracil by the hydride shift mechanism shown in Scheme 18 involving formic acid as a product of formaldehyde oxidation. This is the only prebiotic synthesis of thymine so far described starting from one-carbon atom precursors as simple as formamide and formaldehyde.

The mixture of formamide and formaldehyde is also responsible for the formation of adenine and purine acyclonucleosides **32–34**, probably by a formose-like condensation of an activated formaldehyde on the exocyclic formyl moiety of formylpurine and adenine derivatives **28–29**. Nucleosides

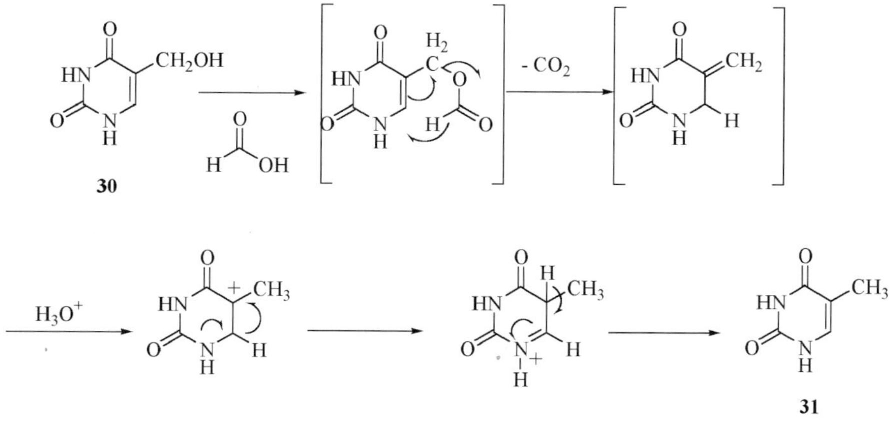

Scheme 18

are usually obtained in low yield and selectivities under prebiotic conditions starting from preformed sugars and heterocyclic bases. With the exception of the one-pot synthesis of purine acyclonucleosides from formamide, no data are available on the synthesis of large panels of nucleoside derivatives from a one-carbon fragment compound as chemical precursor (see next paragraph for additional details on canonical prebiotic syntheses of nucleosides). In accordance with the data reported above on the degradation of 2′-oligodeoxynucleotides by formamide and clays, when the degradation of DNA was performed in the presence of TiO_2 the reactivity of purines decreased while that of pyrimidines increased, showing that catalysts may tune the rate of degradation of polymer-embedded nucleobases.

3 Prebiotic Synthesis of Purine and Pyrimidine Nucleosides

The synthesis of purine and pyrimidine nucleosides is one of the least understood aspects of the prebiotic chemistry of nucleic acids on the primitive Earth. The first attempt to synthesize nucleoside derivatives under simulated primitive Earth conditions was probably that reported by Ponnamperuma and Orgel in 1967 [104]. In this study a mixture of adenine, 2′-deoxyribose and cyanide (CN^-) was UV-irradiated to give a mixture of 2,3-dideoxy-(9-purinyl) pentoses that was not further characterized. In the 1970s, Orgel described a successful prebiotic synthesis of purine nucleosides by heating together in dry phase purine bases and D-ribose. Under these experimental conditions ribose was found to be selectively bound to the exocyclic primary group of adenine **1** and guanine **6** to form the corresponding 6-ribosylamino adenine and 2-ribosylamino guanine derivatives **35–36** (Scheme 19).

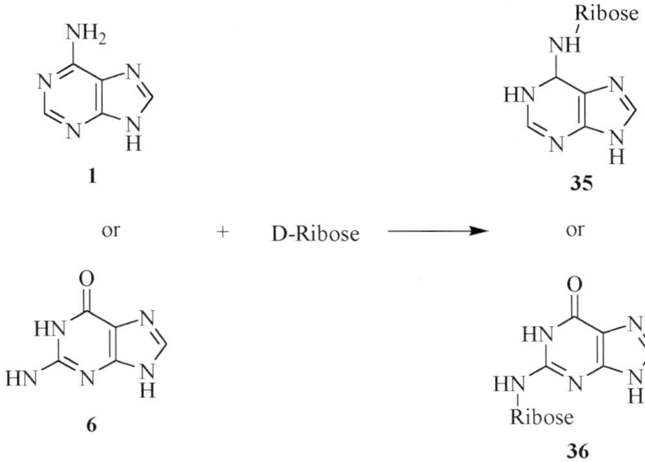

Scheme 19

When the reaction was performed in the presence of a source of Mg^{2+} ions and inorganic polyphosphates, the natural nucleosides β-adenosine **37** and β-guanosine **38** were also formed in 4% and 9% yields, respectively [105]. In a continuation of this study the authors showed that other salts and minerals were able to catalyze this condensation. A mixture of 9-D-ribopyranosylhypoxanthine **39**, α-inosine **41** and β-inosine **40** (ratio c.a. 1 : 1.5 : 3.0, respectively) was for instance obtained upon heating hypoxanthine **2** and D-ribose in solid-state conditions in the presence of minerals including apatite, $CaHPO_4$, $MnSO_4$, $(NH_4)_2SO_4$ and of the salts that remain as residue after the evaporation of seawater (NaCl, $MgCl_2$, $CaCl_2$, and $CaSO_4$) (Scheme 20) [106]. Little or no synthesis occurred in the presence of clays (montmorillonite, kaolinite and zeolite), NaCl and $CaCl_2$ alone.

The anomeric composition of 9-D-ribopyranosylhypoxanthine was not determined while the maximum yield of β-inosine based on the initial hypoxanthine content was 20%. The concentration of ribose was always a fifth of that of hypoxanthine. Similar results were obtained in the case of the condensation process of adenine, guanine and xanthine. β-Adenosine **37** was for instance recovered in a yield of 2.3% in the presence of α-adenosine (1–2%) **42** and 6-ribosylaminopurine **43** as the major product (50–70%) (Scheme 21). Different pyrimidine nucleobases (cytosine, uracil and thymine) failed to produce the corresponding nucleosides when treated with D-ribose at high temperatures. A mixture of adenosine nucleoside derivatives was also obtained after irradiation with a gamma source of adenine and 2′-deoxy-D-ribose in a dry phase. The anomeric composition of these products was not determined. When the reaction was performed under similar experimental conditions in the presence of phosphates (K_2HPO_4, KH_2PO_4 and NaH_2PO_4) products characterized by chromatographic properties similar to those of authentic

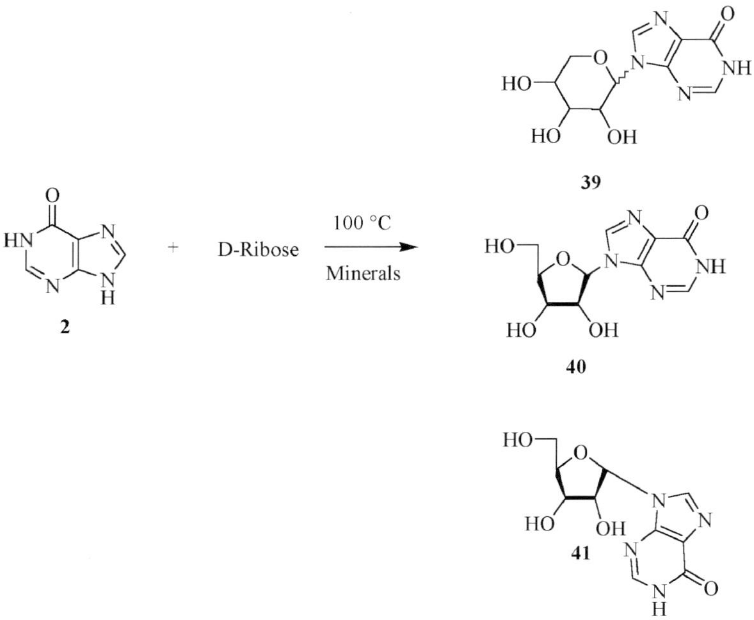

Scheme 20

2′-deoxyadenosine monophosphate were also recovered in the reaction mixture [107].

In the step-by-step synthesis of nucleosides the availability of ribose is not rationalized by a realistic prebiotic pathway. A complex mixture of sugar derivatives, including treose, tetrose, pentose and hexose derivatives of all stereochemistries were obtained heating formamide under basic conditions (formose reaction) in which ribose was estimated to be less than 1% [108, 109]. Further improvements of the formose reaction have been reported performing the condensation with glycolaldehyde phosphate and glyceraldehyde-2-phosphate in the presence of hydrotalcite [110] or using inorganic compounds able to selectively coordinate furanose *cis*-diols [111, 112]. Irrespective of the mechanism of formation, ribose is also characterized by a low stability [113] and by a high reactivity toward nucleophiles [114]. These problems can be overcome under experimental conditions in which ribose is selectively sequestrated in a stable form and successively used for the assembly of the nucleoside scaffold. Recently, Springsteen and Joyce reported that cyanamide reacts selectively with ribose to give a stable ribose-cyanamide adduct which crystallizes spontaneously in aqueous solution [115]. When the reaction was performed with a racemic mixture of D- and L-ribose, enantiomerically twinned crystals were formed that contain discrete homochiral domains. Moreover, ribose-cyanamide adduct reacts with cyanoacetylene to form pyrimidine nucleosides [116, 117]. Thus, the

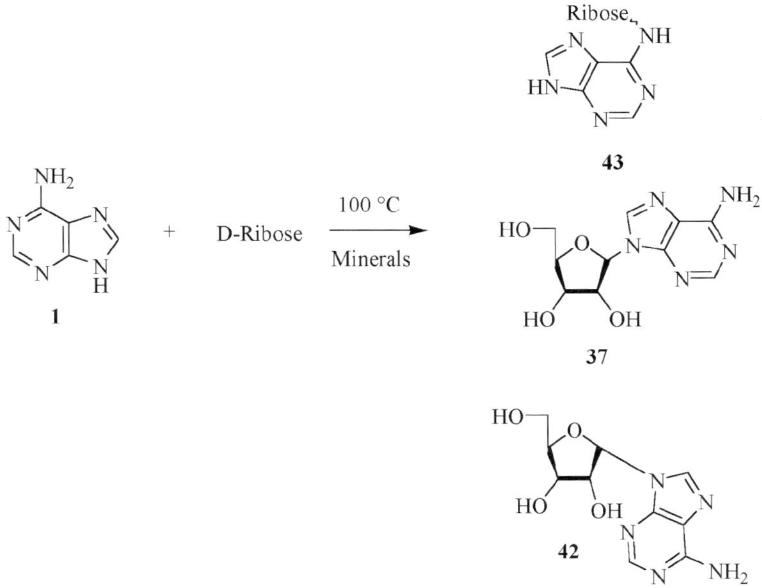

Scheme 21

authors suggested that the ribose-cyanamide adduct can be considered as a solid-phase reservoir of ribose for the prebiotic synthesis of pyrimidine nucleosides.

4 Prebiotic Synthesis of Purine and Pyrimidine Nucleotides

The prebiotic synthesis of nucleotides by phosphorylation of parent nucleosides was studied under different experimental conditions as a potential prebiotic process leading to present-day nucleic acids. This transformation might occur under all three conditions: water solution, solid phase or wet-dry cycles. In the 1960s, Ponnamperuma described the phosphorylation of nucleosides in dry phase at 160 °C with either phosphoric acid (H_3PO_4) or the monobasic salts NH_4HPO_4, $NH_4H_2PO_4$ and $Ca(H_2PO_4)_2$. Some of these orthophosphates converted easily to condensed phosphates [118]. When this reaction was performed in solution condensing agents played an important role. Uridine-5′-monophosphate could be for instance obtained from uridine and inorganic phosphate in aqueous solution in the presence of cyanogens, cyanoformamide, cyanate, cyanamide, thioformate and others. The yields were always low even when a large excess of condensing agent was used [119]. Formamide was also used as condensing agent during the phosphorylation process. The phosphorylation of nucleosides and deoxynucleosides with ammonium or alkali dihydrogen phosphates in formamide was originally reported by Schoffstall [120, 121]. The reaction was studied in a wide range of temperature (25–140 °C) and reaction times (12 hrs–70 months). The conditions necessary in formamide were found to be milder than those required to perform the phosphorylation under solid-state conditions [122].

The selectivity of phosphorylation generally favoured the functionalization at the 5′-position of ribose or 2′-deoxyribose, probably because of the higher reactivity of the primary hydroxyl group with respect to the secondary hydroxyl groups on the 2′- and 3′-positions. As an example, treatment of adenosine **42** with KH_2PO_4 at 37 °C for 6.5 months afforded adenosine-5′-monophosphate **44** in 7.6% yield as the main product in addition to adenosine-2′-monophosphate (2′-AMP) **45** (3.4%), adenosine-3′-monophosphate (3′-AMP) **46** (1.4%) and adenosine 2′,3′-cyclic phosphate (2′, 3′-cAMP, 0.1%) **47** (Scheme 22) [123]. Cyclic nucleosides were found to be more abundant for reactions performed at 125 °C or higher. As for the 2′-deoxynucleosides, thymidine showed a preference for the formation of the corresponding nucleotide 5′-monophosphate, while 2′-deoxyadenosine preferentially formed the nucleotide-3′-monophosphate. Noteworthy, in these transformations formamide is instrumental for the solubilization of both nucleosides and inorganic phosphates and acts as a catalyst for the esterification processes [124]. In addition, data on the hydrolysis of nucleosides and on the isomerization to nucleotide isomers showed that formamide could perform both phosphorylation and dephosphorylation. Cyclic nucleosides were found to resist to formamide hydrolysis. Conclusions reached from mechanistic studies show that the phosphorylations are a series of equilibria, with cyclic nucleotides being formed irreversibly. The phosphorylation of nucleosides in the presence of formamide was also obtained with ammonium phosphate at 37 °C [125].

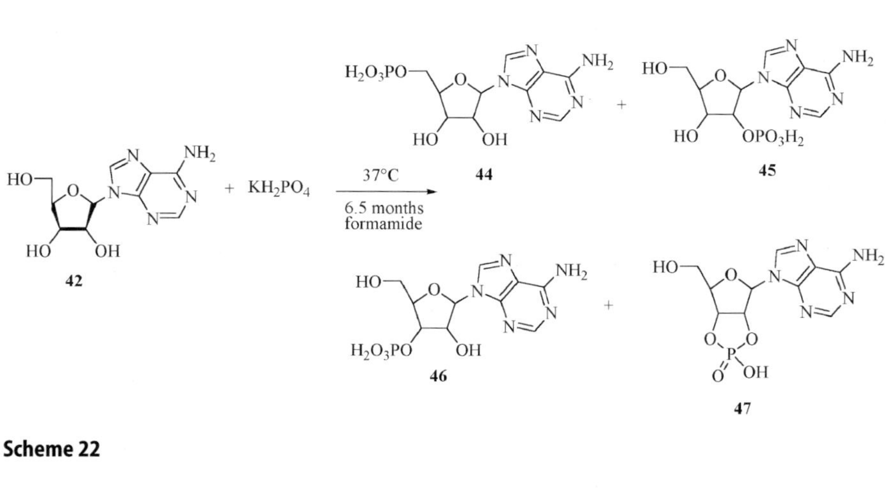

Scheme 22

Urea is another prebiotic condensing agent for the phosphorylation of nucleosides. Condensed phosphates have been obtained from ammonium phosphate and urea [126], while alcohols are easily phosphorylated at 150 °C by urea and phosphoric acid [127]. When an equimolar mixture of uridine 48 and Na_2HPO_4 was heated at 100 °C in the presence of a ten-fold molar excess of urea and ammonium chloride, 97% of the phosphate was found to be incorporated into the nucleoside moiety mainly as uridine 5′-monophosphate 49 and uridine 2′,3′- cyclic phosphate 50 (Scheme 23).

In the absence of urea very little phosphorylation occurred. The phosphorylation of cytidine under similar experimental conditions afforded the incorporation of 96% of the available phosphate, while adenosine and guanosine incorporated only 80% and 71% of the phosphate, respectively. The major product of the phosphorylation of nucleosides was the nucleotide 2′,3′-cyclic phosphate, although nucleotide-5′-phosphate was also observed in good yield. Thymidine originated large amounts of both 3′- and 5′-phosphates. In each case some carbamylation occurred on the sugar and on the nucleobases as a side-reaction. Only a low yield of oligonucleotides was obtained in the reaction mixture [128]. Alternatively, heating uridine in the presence of apatite, ammonium oxalate and additional reagents such as cyanamide, urea and imidazole or cyanate resulted in the formation in 10–20% yields of the 2′-3′ and 5′-isomers of uridine monophosphate [129]. 2′,3′-Cyclic phosphates can be used as monomers for the synthesis of oligonucleotides in dry phase. When for instance adenosine 2′,3′-cyclic phosphate was heated with 1,2-diaminoethane under anhydrous conditions, oligoadenylates were obtained in acceptable yield [130]. In this non-templated polymerization the natural 3,5′-linkages were found to be more abundant that the 2′,5′-linkages (ratio 3,5′- *versus* 2,5′-linkages about 2:1). The same reaction performed in aqueous solution in the presence of a template was less efficient and afforded a dimer as the major reaction product which was characterized by 97% of 2′,5′-linkage [131–133]. The preference for the 2′,5′-linkage appears to be due to the optimal geometry of the transition state for the reaction so that lower energy was associated to the preferential formation of this

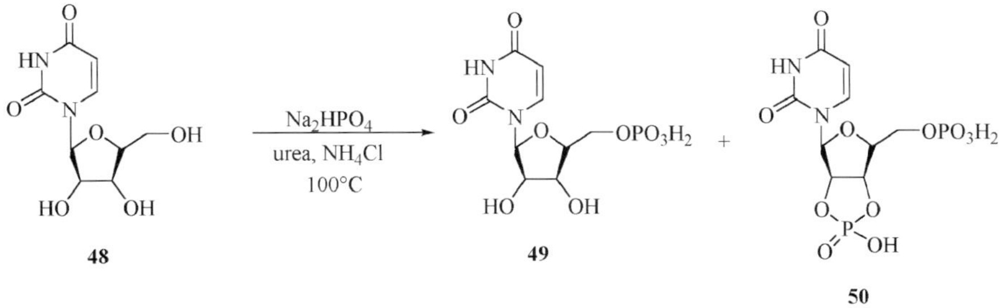

Scheme 23

specific bond [134]. On the other hand, oligonucleotides bearing non natural 2, 5′-linkages are less stable than those with 3′, 5′-linkages [135]. Thus, from a thermodynamic point of view repeated cycles of synthesis and hydrolysis of heterogeneous oligonucleotides will eventually lead to the formation of present day 3′,5′-linked nucleic acid derivatives. A re-assessment of the use of urea as condensing agent was made by Zubay [136]. In this study, a procedure modified relative to previous studies was applied, which included lower reaction temperatures, the use of a mixture of monohydrogen and dihydrogen phosphates instead of just the former and the elimination from the reaction mixture of some of the original components, such as NH_4HCO_3 and NH_4Cl. Reactions were then performed in dry phase, using different amounts of urea, inorganic phosphate and a range of temperatures between 55 and 95 °C. Under these experimental conditions 5′-monophosphate nucleotides were recovered as the main reaction products. The 2′,3′-cyclic phosphate nucleotides become the major products only under prolonged reaction times and for temperatures higher than 100 °C. Under hydrolytic conditions the removal of phosphate from the 3′-OH moiety was preferred over that on the 5′-position. Trimetaphosphate, polyphosphate and hydroxylapatite were also studied as phosphate donors. Polyphosphate was found to be almost as effective as orthophosphate and to have the same requirement for urea. Hydroxylapatite was only 10% as effective as orthophosphate. Trimetaphosphate was ineffective. Some organophosphates were also analyzed as phosphorylating agents showing a reactivity similar to orthophosphate.

In the 1960s, Schwartz described the phosphorylation of adenosine with trimetaphosphate to yield 2′- and 3′-AMP. The strong alkaline conditions used for this transformation were not likely to occur on the primitive Earth [137]. Similarly, all natural ribonucleosides were phosphorylated to corresponding 2′- and 3′-nucleotide monophosphates with sodium trimetaphosphate at high pH and temperature [138, 139]. When the reaction was performed under similar experimental conditions at lower pH, 2′,3′-cyclic phosphate nucleotides were recovered as the major products [140]. Magnesium ion catalyzes this transformation in neutral water solution [141].

Recently, Zhao has described the phosphorylation of adenosine **42** with trimetaphosphate by a wet-dry cycle process performed at neutral pH. Metal ions were able to catalyze the reaction, Ni(II) being the most active catalyst (30% yield of phosphorylated products, twice the yield produced by magnesium ion). 2′,3′-Cyclic AMP **47** (10%) and 5′-adenosinetriphosphate 5′-ATP **50** (13.0%) were obtained as the main reaction products (Scheme 24) [142]. Different prebiotic syntheses are reported for 5′-ATP [143–149]. 5′-ATP **50** has been for instance synthesized by the phosphorylation of 5′-ADP **51** in aqueous solution containing cyanate and insoluble calcium phosphate (Scheme 25) [150]. Similarly, 5′-ADP was synthesized from 5′-AMP. The yield

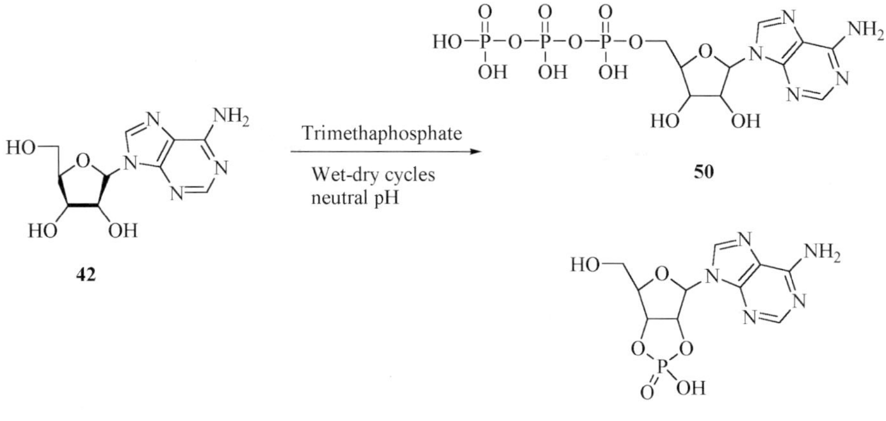

42

50

47

Scheme 24

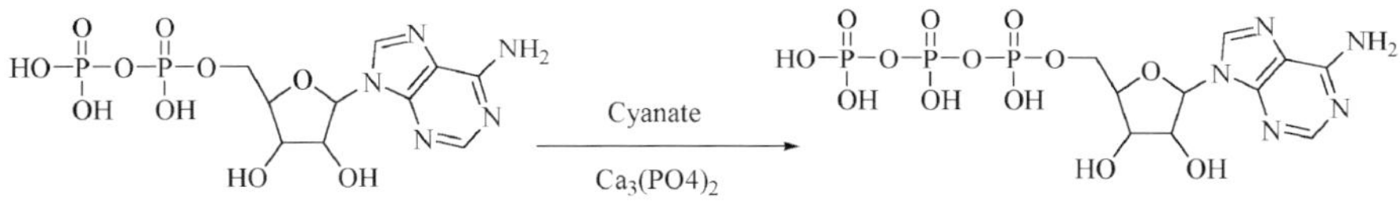

51

50

Scheme 25

of phosphorylated products was correlated with the adsorption of substrates on the calcium phosphate surface.

In the 1970s, Ferris described the effect of diiminosuccinonitrile (DISN) as a condensing agent. DISN is formed by oxidation of diaminomaleonitrile (DAMN) and it is recognized as an important intermediate in the polymerization of HCN [151]. The formation of DISN is catalyzed by metal ions, such as the non-exchangeable Fe^{3+} ions present in the clay lattice [152, 153]. Even if DISN was not an effective condensing agent in the conversion of nucleosides to nucleotides with inorganic phosphate in aqueous solution, it nevertheless efficiently catalyzed the cyclization of 3′-adenosine monophosphate to adenosine 2′,3′-cyclic phosphate. Thus, when adenosine 3′-monophosphate **46** was heated at 60 °C in the presence of DISN and Mn^{2+}, a 39% yield of adenosine 2′,3′-cyclic phosphate **47** was obtained (Scheme 26) [154].

The reaction probably proceeds through the formation of an imino analogue of mixed anhydrides by nucleophilic displacement of cyanide during the phosphate activation. Cyanogen bromide was found to be more effective than DISN in this transformation. Since higher yields were obtained in the preparation of the cyclic phosphate from 3′-adenosine phosphate than from 2′-adenosine phosphate, the authors suggested the formation of an activated

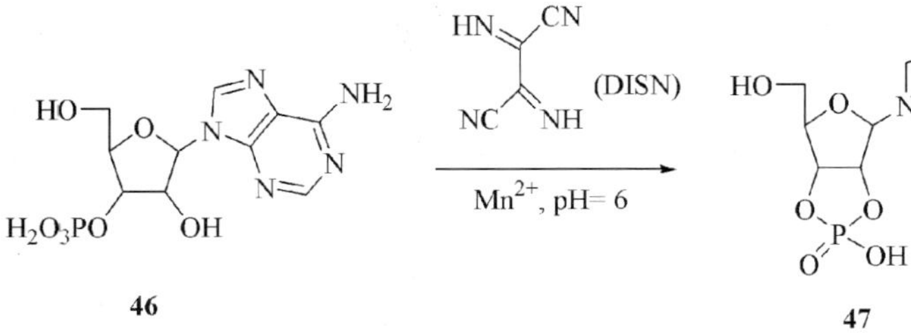

Scheme 26

phosphate adduct between the 3′-phosphate group, DISN and metal ions as a key intermediate of the reaction. The DISN-mediated cyclization of 3′-AMP was also catalyzed by Zn^{2+} homoionic montmorillonite clays.

5 Prebiotic Synthesis of Oligonucleotides

The first attempts of prebiotic synthesis of oligonucleotides were mainly directed to design the optimal conditions for the formation of the phosphate ester bond between preformed nucleotide derivatives. In the 1960s, Schramm proposed polymetaphosphate (PMP) as a condensing agent for both nucleosides and nucleotides [155, 156]. Although the polymer obtained from PMP showed some similarities to oligonucleotides of biotic origin, such as a similar coding ability in a cell-free system of *Escherichia coli* and hypochromicity effect in the presence of polyadenylates, further experiments suggested the presence in their structure of a large amount of non-natural linkages as well as branching and cross-linking between chains [157–159]. In addition, PMP requires anhydrous conditions and lacks geological relevance. Also polyphosphoric acid and its salts, produced by heating ammonium hydrogen phosphates at high temperatures, have been used as condensing agents in the preparation of short (4- or 5-mers) polycytidylates, polyadenylates and heterogeneous oligonucleotides [160]. A series of degradation experiments of these products performed with alkaline phosphatase from *Escherichia coli* showed that only a single 2′(3′)-phosphomonoester bond was present per chain, suggesting that a significant fraction of the product was closely related to natural oligonucleotides. In principle, the prebiotic synthesis of the oligonucleotide chain can be performed more easily in the presence of a template. In this case, molecular recognition processes based on specific hydrogen bond interactions between complementary nucleobases can direct the monomer in the correct spatial position to form the phosphodiester bond with low energy requirement. Based on this model, Orgel described the first selective polymerization of adenosine-5′-monophosphate (5′-AMP) and guanosine-5′-monophosphate (5′-GMP) in the presence of complementary oligonucleotides as templates (polyuridylic and polycytidylic acids, respectively) using different condensing agents [161–163]. 2′,5′-Phosphodiester linkages were predominantly formed. These experiments marked an important advancement in the prebiotic synthesis of oligonucleotides, introducing the use of activated nucleotides, mainly on the 5′-position of the sugar, for the polymerization process. In the 1970s, Orgel described the first example of an efficient prebiotic synthesis of oligonucleotides using together the template and the activated nucleotides approaches [164, 165]. The phosphate activating group, that was usually a good leaving group such as imidazole and 2-methylimidazole, played a relevant role in driving the polymerization toward longer oligomers [166]. Under these experimental conditions, oligonucleotides with chain lengths not exceeding 20-mers were obtained. Longer products were not isolated in appreciable amounts because of the competing formation of the 2′,5′-linkages [167, 168]. In this context, imidazolide derivatives of ribonucleotide diphosphates (examples of their structure are reported below) were used for the polymerization of oligonucleotides linked by pyrophosphate bonds [169, 170]. Oligomers produced by these processes could be replicated and elongated by template-directed syntheses [171–173]. Noteworthy, the prebiotic synthesis of oligonucleotides in the presence of a template can be selectively catalyzed by minerals which were widely diffused on the primitive Earth, such as montmorillonite clays [174]. Montmorillonites, which are also able to catalyze the prebiotic synthesis of nucleobases, may have served both to concentrate nucleotides (or their activated derivatives) from dilute aqueous solutions and as catalysts for their polymerization. Ferris described several examples of montmorillonite catalyzed synthesis of oligonucleotides [175–179]. Linear and cyclic oligonucleotides containing both 3′,5′- and 2′,5′-phosphodiester bonds, and pyrophosphate bonds ($N^{5'}$ppN bonds), were for instance synthesized in acceptable yield by condensation of 5′-phosphorimidazolide cytidine (ImpC) on montmorillonite (Fig. 2).

The prebiotic relevance of these heterogeneous oligonucleotides was shown by using them as templates for the preparation of the complementary oligoguanylates from 5′-phosphorimidazolide of guanidine (ImpG)

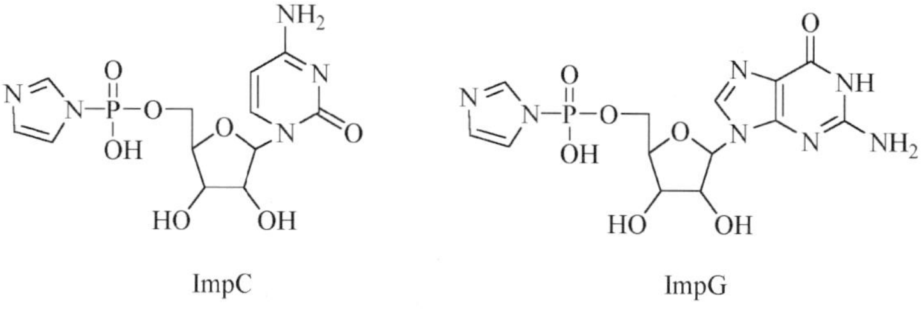

Fig. 2 Structures of 5′-phosphorimidazolide cytosine and hypoxanthine derivatives

and homoionic montmorillonite (that is a montmorillonite characterized by the presence of only one type of cation) (Fig. 2). The experiments performed in the absence of templates yielded only short oligonucleotides [180]. A detailed investigation of the effect of the structure of the heterocyclic phosphate activating group on the synthesis of oligonucleotides on montmorillonites showed that aminopyridine derivatives, such as the 4-dimethylaminopyridine, were more effective than imidazoles for the preparation of oligoadenylates [181]. This reactivity was mainly correlated to the pKa value of the heterocyclic activating group (the most effective pKa values were in the range of 6.0–9.0) and to its ability to stabilize the newly formed charges during the nucleophilic displacement by the hydroxyl moiety. Since only a few prebiotic syntheses have been described for imidazole and aminopyridine derivatives, the most common heterocyclic compounds present on the primitive Earth were analyzed as possible activating groups. Among them, special attention was devoted to adenine and to some adenine derivatives [182]. Activated nucleotides of 1-methyladenine (pKa 7.2), 3-methyladenine (pKa 6.1), 2-methyladenine (pKa 5.1) and adenine (pKa 4.12) were prepared by reaction with the appropriate 5′-NMP in the presence of a condensing agent (the structures of the activated adenosine nucleotides are reported in Fig. 3). The polymerization was carried out at pH 8 in the presence of sodium homoionic montmorillonite to give oligomers containing up to eleven monomeric units. Control reactions performed in the absence of montmorillonite yielded only dimers. Adenine derivatives showed

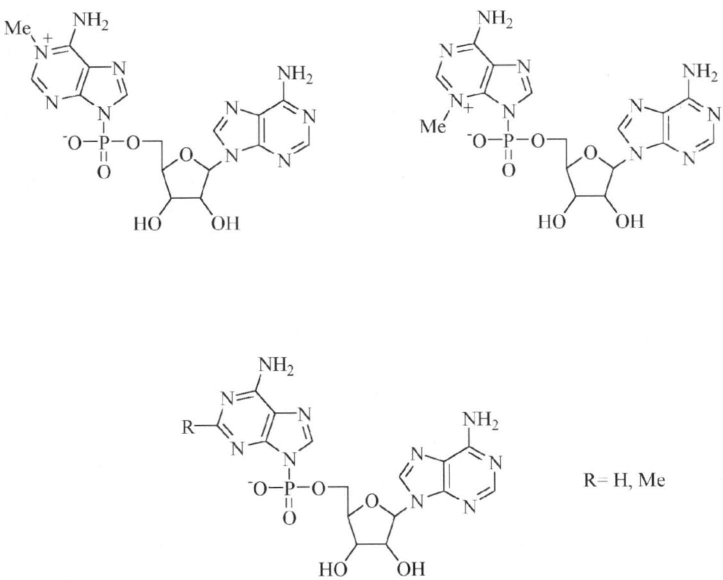

Fig. 3 Examples of adenine-5′-phosphate activated nucleotides

a reactivity similar to imidazoles as phosphate activating groups. Moreover, adenine derivatives whose pKa values are between 6 and 9 yielded longer oligomers than those whose pKa values are < 6. The yield of oligomers was not enhanced by lowering the pH of the reaction. The activated nucleotides containing the methyladenine activating group yielded longer oligonucleotides and higher percentage of natural 3′,5′-phosphodiester bonds, as evaluated by bacterial alkaline phosphatase (APH) and ribonuclease T_2 (Rnase T_2) treatment of the purified products. This adenine derivative was found to be second only to 4-dimethylaminopyridine in its efficiency in forming long oligomers. Appreciable formation of 2′, 5′-phosphodiester bond, of pyrophosphate bond and of cyclic oligonucleotides was also observed.

In agreement with the higher reaction rate observed for monomers carrying an activated phosphate, relative to differently activated monomers, the percentage of the pyrophosphate bond embedded into products increased with the increase of the oligomer molecular weight [183, 184]. The montmorillonite catalyzed reaction has been extended to the phosphorimidazolides of guanosine, inosine and uridine [185]. The yield of oligomers was enhanced by the presence of $MgCl_2$ in the reaction mixture because of its effect on the binding of activated nucleosides to the montmorillonite surface [186]. The rate constants for the polymerization of pyrimidine nucleotides on montmorillonite were higher than those of purine nucleotides [188]. Hydrothermal environments present on the sea floor of the primitive Earth might also have served as a local reactor for the synthesis of oligonucleotides from nucleotides [189]. These environments are characterized by the presence of a constant temperature gradient between the hot vents and their surroundings, where synthetic and degradative pathways can be operative [190]. In this context, the interface between the hot and the cold region deserves special attention. In fact, in a scenario in which the mass transfer across the interface is faster than the degradative processes, products obtained at high temperatures may survive long enough to be used in the molecular evolution process. Matsuno reported the first example of synthesis of oligonucleotides from nucleotide monophosphates in the absence of condensing agents in a simulated hydrothermal environment [191]. Similar results were obtained for the preparation of peptides from amino acids [192]. When adenosine monophosphate and $ZnCl_2$ were introduced in a flow reactor in which a high temperature (110 °C) and high pressure (13 Mpa) fluid was injected into a low temperature (0 °C) chamber maintained at the same pressure as the fluid, both 3′,5′ and 2′,5′ dimers and uncharacterized trimers were detected in the reaction mixture. This example sets one of the most plausible scenario for the prebiotic synthesis of oligonucleotides under high temperature conditions.

The polymerization of activated nucleotide derivatives has been also performed under eutectic phases in ice, a special microenvironment providing both concentration of reagents and preservation of newly formed oligomers. For example, Kanavarioti and Monnard reported the polymerization of polyuridylates (up to 11-mer) in frozen samples of phosphoimidazole derivatives of uracil in the presence of Mg(II) and Pb(II) ions [193]. Similar results were obtained in the polymerization of mixtures of purine and pyrimidine phosphoimidazole derivatives to obtain heteropolymers of medium-size (at least 17 units long) [194]. These oligomers were characterized by approximately 40% of natural 3′-5′ linkages in contrast to previous values (< 10%) found in solution experiments [195, 196]. The nucleobase content in heteropolymer varied according to the initial concentration of the activated monomers in the reaction mixture.

6 Pre-RNA and Pre-DNA Molecules

Other potentially informational systems have been discovered, all closely related to nucleic acids. A thorough analysis of the properties of nucleic acid analogues showed that possibilities exist other than DNA or RNA. Nucleic acid alternatives and the chemical properties fundamental for their biological functions have been systematically investigated by Eschenmoser [197]. Representative examples of these nucleic acid alternatives are 2′,3′-dideoxy-glucopyranosyl-(6′-4′)-β-oligonucleotides (homo-DNA) [198–201], a family of diastereomeric (6′-4′)-hexopyranosyl-oligonucleotides derived from D-allose, D-altrose and D-glucose, (4′-2′)-pentopyranosyl-oligonucleotides derived from D-ribose (pyranosyl-RNA) [202, 203], L-lyxose [204], D-xylose [205] and L-arabinose [206], (4′-3′)-pentopyranosyl-oligonucleotides derived from D-ribose and L-lyxose [207] and finally L-α-threofuranosyl-(3′-2′)-oligonucleotides (TNA) (Fig. 4) [208, 209]. TNA shows a structure simpler than RNA characterized by four instead of five carbon atoms in the sugar moiety and only three instead of six possible regioisomeric phosphodiester linkages. This RNA analogue has the potential to serve as a template in non-enzymatic template-directed synthesis of RNA from activated nucleotides.

The property of nucleobase-pairing of these nucleic acid alternatives has been analyzed measuring the melting point temperature (T_m) for selected duplexes of self-complementary dodecamer sequences. Low efficiency base-pairing properties were observed in the hexopyranosyl series (with the exception of the homo-DNA oligonucleotides), while very strong pairing was measured for the pentapyranosyl and TNA oligonucleotide series, the pairing capability of TNA being comparable in strength to that of RNA. In this latter case, the base-pairing with antiparallel backbone orientation was favoured over base-pairing with parallel backbone orientation. Moreover, TNA was capable of cross-pairing with both RNA and DNA showing to be more stable toward hydrolytic strand fission than RNA [210]. Replacement of adenine by 2,6-diaminopurine in TNA base sequences resulted in a marked enhancement of duplex stability [211]. TNA analogues produced using both 3′-NH- and

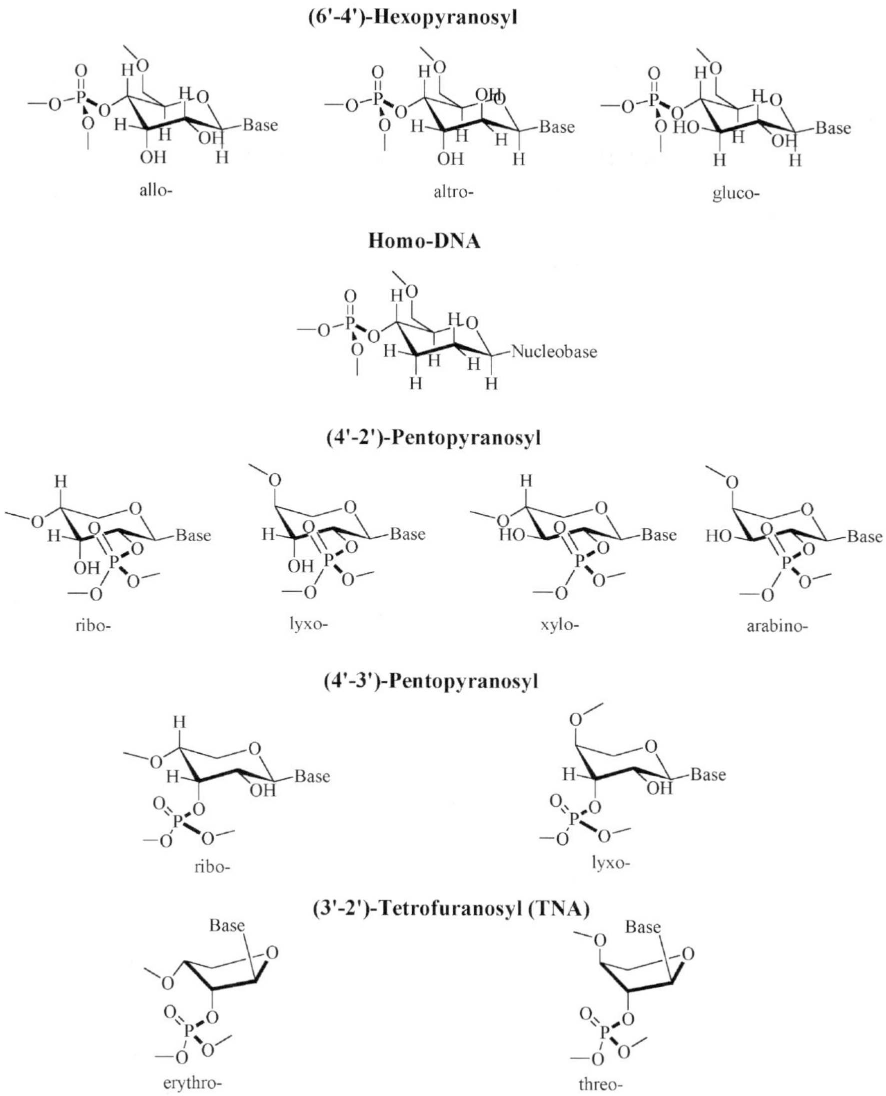

Fig. 4 Structure of nucleic acid alternatives

2′-NH-nitrogenous tetrose derivatives, which can be formed by aldolization of glycolaldehyde and ammonia under prebiotic conditions, showed pairing and cross-pairing properties similar to those of TNA [212]. The presence of amino groups on the sugar moiety opens the possibility for linkages alternative to phosphodiester bonds, such as the guanidinium and amide bonds, more stable toward hydrolysis (Fig. 5) [213, 214].

Since *C*-nucleosides, in which the heterocyclic moiety is bound to sugar by a C – C bond, are more stable than *N*-nucleosides, the relevance

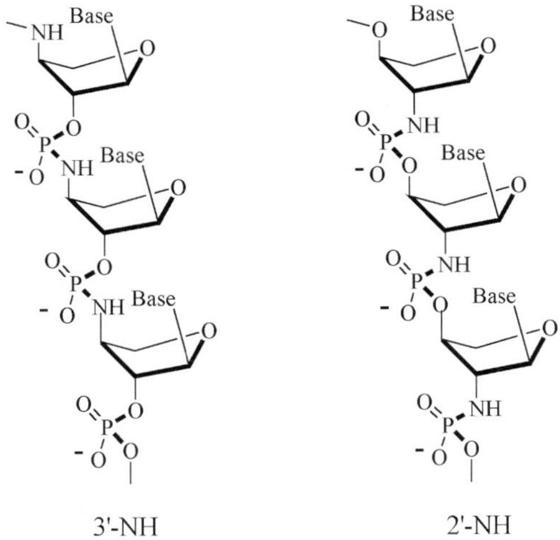

Fig. 5 3′-NH- and 2′-NH-phosphoramidate sequences of TNA

of a *C*-nucleosidation process in the etiology of TNA-analogues has also been investigated. The synthesis of *C*-nucleoside derivatives was for instance easily performed by reacting the purinoid compound 5,8-dicarba-2,6-diaminopurine with pyrroline, threose and ribose derivatives [215]. The role of these analogues on the prebiotic synthesis of nucleic acid alternatives is being analyzed.

Noteworthy, some oligonucleotide analogues are able to replicate in a template directed synthesis without significant enantiomeric cross-inhibition. For example, Kozlov, Orgel and co-workers described that cytidinyl hexitol nucleic acids (HNAs) can be used for the efficient oligomerization of D-guanosine 5′-phosphoro-2-methylimidazole (D-2-MeImpG) in the presence of L-guanosine 5′-phosphoro-2-methylimidazole (L-2-MeImpG), a compound that usually acts as an inhibitor during the oligomerization with DNA and RNA templates [216]. These results are explained supposing that L-2-MeImpG is excluded from the "active site" of polymerization by a competition with the right isomer D-2-MeImpG or that it is confined in a specific conformation unable to afford covalent bond formation. Altritol nucleic acids (ANAs) showed a similar behaviour and were found more efficient templates than DNA, RNA and HNA during the nonenzymatic polymerization of 5′-phosphoro-2-methyl imidazolides [217].

Peptide nucleic acid (PNA) is another analogue of RNA and DNA that has been considered as a potential ancestor of present day nucleic acids. In this molecule the natural sugar-phosphate backbone has been replaced by peptide-like linkages [218]. In recent years, novel syntheses of PNA have been reported mainly focused on their application for antisense and antigene therapies [219]. The physical-chemical properties of PNA make it both

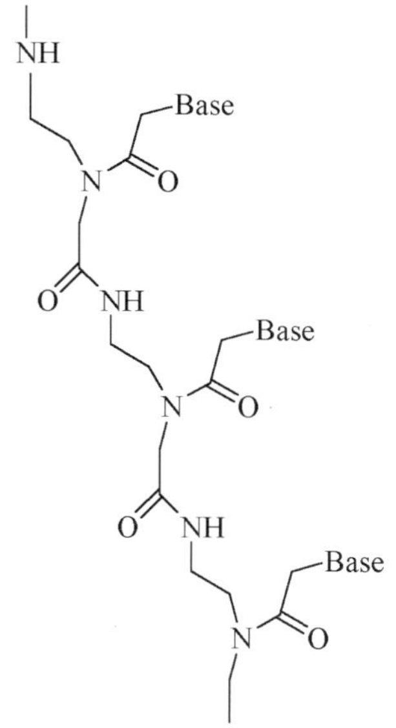

Fig. 6 Structure of PNA

a candidate for the origin-of-life related development of informational self-replicating polymers and a flexible important tool for nanotechnology and innovative analytical and therapeutical approaches. PNA is an alternative nucleic acid having a pseudopeptide uncharged and achiral backbone that makes it very stable in biological fluids. The pseudopeptide backbone is composed of N-(2-amino-ethyl)glycine AEG units to which the nucleobases are linked through methylene carbonyl linkers (Fig. 6) [220]. The very fact that a nucleobases polymer forms that is endowed of exactly the same informational capacity as DNA or RNA but that is made of 4 different atoms (H, C, N, O) instead of 5 (lacking P) has suggested its possible role in early origin-of-life scenarios [221]. However, the likelihood of this role has met the objection that PNA monomers cyclize when they are activated, making oligomer formation unlikely under prebiotic conditions.

Miller reported a plausible prebiotic synthesis of the PNA monomers, which are ethylendiamine (ED), AEG and purine and pyrimidine acetic acid derivatives [222]. Spark discharge experiments with a mixture of CH_4, NH_3, N_2 and H_2O, using both high and room temperature apparatuses, afforded a low amount of ED and AEG. A better result was obtained for the synthesis of ED by NH_4CN polymerization in the presence of H_2CO. Scheme 27 shows the mechanism hypothesized for the production of ED from ammonium cynide and HCN trimer.

AEG might be obtained via the Strecker synthesis (Scheme 28).

In spite of these possible initial difficulties, the interest of PNAs as informational polymers remains high. Two relevant properties in this area have

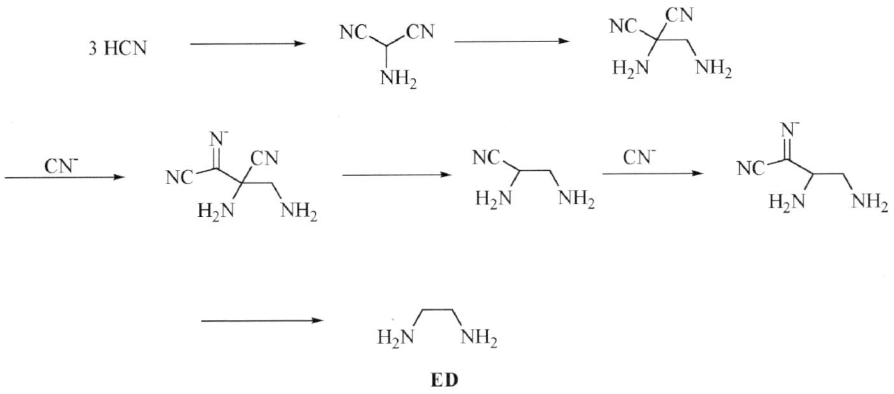

Scheme 27

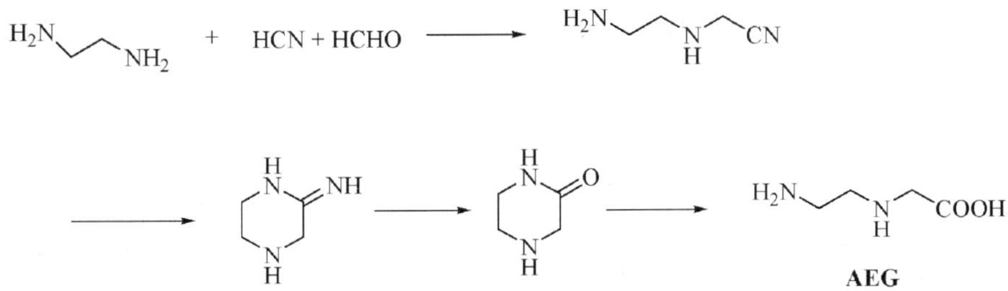

Scheme 28

been described: the information transfer from DNA to PNA by template-directed syntheses [223], and the template switching between PNA and RNA oligonucleotides [224]. Homopyrimidine PNA binding to double stranded DNA results in a triplex invasion complex in which one PNA binds to the homopurine DNA stretch by ordinary Watson–Crick base-pairing, and another PNA binds to the PNA-DNA duplex by Hoogsteen binding. This leaves the non-complementary strand unpaired, forming a structure termed P-loop [225]. Due to its achiral nature, PNA/DNA duplexes can be both parallel and antiparallel. The properties of P-loops have been extensively studied, among which the ability to initiate transcription *in vitro* [226]. PNAs have shown high versatility in their interaction with nucleic acids: PNA-based structures may function as artificial primosomes [227], may tailor the activity of restriction endonucleases [228], form oligonucleotide-PNA-chimaeras [229], and interfere with methylation enzymes [229]. The problems associated with the cellular delivery of PNA, and the possible solutions, have been reviewed [230].

Oligonucleotides bearing acyclic sugar moieties in their structure were also suggested as plausible pre-RNA and pre-DNA molecules. Acyclonucleosides, such as glyceronucleosides, were studied as achiral monomers for the preparation of some "flexible" oligonuleotides analogues with potentially limited problems of enantiomeric cross-inhibition. Unfortunately, Schneider and Benner showed that the incorporation of glyceronucleosides into oligonucleotide analogues decreases the melting temperatures of duplexes [231].

7
Concluding Remarks

Although a clear scenario for the prebiotic emergence of genetic informational polymers is not yet available, the 50 years that followed the seminal Urey–Miller experiment have provided important answers to the majority of the possible facets of the problem. What is actually still lacking is a unitary chemical frame: if polymers had to form, the system had to contain all the necessary starting components. If it is true that life is a robust phenomenon, as it appears from its early onset, the polymerizations had to occur based on a single unitary process and starting from a simple precursor. In principle, the ideal prebiotic precursor is a one-carbon fragment compound that is able to contemporarily synthesize nucleobases and sugars, acting also as condensing agent for the phosphorylation process. The formamide-based efficient production of large panels of both purines and pyrimidines nucleobases and acyclonucleosides, combined with its ability to catalyze the phosphorylation of nucleosides, could provide one plausible solution to the problem of global availabilities. The great versatility of synthesis afforded by common and prebiotically available catalysts adds to this plausibility.

In a prebiotic perspective, what is the reason why nucleobases, possibly in their nucleoside forms, would polymerize and, following their very combination into linear macromolecules, give rise to next level of informational complexity? In the words of Watson and Crick referred to DNA: "it has not escaped our notice" that this type of informational macromolecule and its structural property of complementary specific base pairing "immediately suggests a possible copying mechanism for the genetic material" [232]. The prebiotic perspective changes into a biotic one if a molecular Darwinian selection process comes into play. The selective advantage is intrinsic in the physico-chemical attributes of the system: we suggest that it simply consists of the increased survival of the components upon polymerization. An extensive and unitary analysis of this fact has not been performed. However, numerous observations have been reported since: the rate of hydrolysis of free deoxynucleosides is 10–50 times higher relative to the rate of cleavage of N-glycosyl bonds in single-stranded DNA [233]; the rate of hydrolysis of N-glycosyl bonds in deoxynucleosides is higher that in deoxynucleotides, which is higher than in DNA [234]; depurination is 4 times faster in single- versus double-stranded DNA [235]. Although performed in different systems and with different techniques, taken together these observations indicate that upon polymerization the stability of the starting monomers changes. Whether this fact is sufficient to provide a Darwinian advantage for the survival of polymers versus the survival of monomers or if other factors (i.e. mass effects, favoured adsorption/protection on minerals, etc.) come into play remains to be established.

Acknowledgements We are grateful to Italian MIUR COFIN 2003 "la catalisi dei metalli di transizione nello sviluppo di strategie sintetiche innovative di eterocicli" for financial support.

References

1. Orgel LE (1994) Scientific American 10:53
2. Eschenmoser A, Lowenthal E (1992) Chem Soc Rev 1
3. Ernst WG (1983) Earth's earliest biosphere. Princeton University Press, Princeton, NJ, pp 41–52
4. Chyba CF, Thomas PJ, Brookshaw L, Sagan C (1990) Science 249:366
5. Kasting JF (1993) Science 259:920
6. Miller S (1987) In: Proceedings of Cold Spring Harbor Symposia on Quantitative Biology, vol. LII 17
7. Wächtershäuser G (1988) Microbiol Rev 52:452
8. Huber C, Wächtershäuser G (1997) Science 276:245
9. Crabtree RH (1997) Science 276:222
10. Cronin JR (1989) Adv Space Res 9:54
11. Anders E(1989) Nature 342:255
12. Maurette M, Jèhanno C, Robin E, Hammer C (1987) Nature 328:699
13. Drobner E, Huber H, Wächtershäuser G, Rose D, Stetter KO (1990) Nature 346:742
14. Hafenbradl D, Keller M, Wächtershäuser G, Stetter KO (1995) Tetrahedron Lett 36:5179
15. Keller M, Blöchl E, Wächtershäuser G, Stetter KO (1994) Nature 368:836
16. Oparin AI (1938) In: The origin of life. MacMillan, New York
17. Saladino R, Crestini C, Costanzo G, Di Mauro E (2004) Curr Org Chem 8:1425
18. Orgel LE (2004) Critical Rev Biochem Molecular Biology 123:39
19. Sutherland JD, Whitfield JN (1997) Tetrahedron 53:11493
20. Maurel MC, Dècout JL (1999) Tetrahedron 55:3141
21. Abelson PH (1966) Proc Natl Acad Sci 55:1365
22. McCord TB, Carlson RW, Smythe WD, Hansen GB, Clark RN, Hibbitts CA, Fanale FP, Granahan JC, Segua M, Matson DL, Johnson TV, Martin PD (1997) Science 278:271
23. Bar-Nun A, Tauber ME (1972) Space Life Sci 3:254
24. Hubbard JS, Voecks GE, Hobby GL, Ferris JP, Williams EA, Nicodem DE (1975) J Mol Evol 5:223
25. Ferris JP, Chen CT (1975) Nature 258:587
26. Sanchez R, Ferris JP, Orgel LE (1966) Science 154:784
27. Krebaum LJ (1966) J Org Chem 31:4103
28. Hirose Y, Ohmuro K, Saigoh M, Nakayama T, Yamagata Y (1991) Origin Life Evol Biosphere 20:471

29. Miyakawa S, Cleaves HJ, Miller SL (2002) Origin Life Evol Biosphere 32:195
30. Ferris JP, Zamek OS, Altbuch AM, Freiman HJ (1974) J Mol Evol 3:301
31. Ferris JP, Hagan WJ (1984) Tetrahedron 40:1093
32. Rank MD, Townes CH, Welch WJ (1971) Science 1083
33. Mumma MJ, Weissman PR, Stern SA (1993) In: Levy EH, Lunine JI (eds) Protostars and Planets III. University of Arizona Press, Tucson, AZ, pp 1177–1252
34. Magge-Sauer K, Mumma MJ, DiSanti MA, Russo ND, Rettig TW (1999) Icarus 90:203
35. Irvine WM (1999) Space Sci Rev 90:203
36. Hidayat T, Marten A, Bèzard B, Ganter D, Owen T, Matthews HE, Paubert G (1997) Icarus 126:170
37. Strecker A (1850) Liebigs Ann Chem 75:27
38. Miller SL (1953) Science 117:528
39. Miller SL (1955) J Am Chem Soc 77:2351
40. Orò J (1960) Biochim Biophys Res Commun 2:407
41. Orò J, Kimball AP (1961) Arch Biochem Biophys 94:217
42. Orò J, Kimball AP (1962) Arch Biochem Biophys 96:293
43. Ferris JP, Orgel LE (1965) J Am Chem Soc 87:4976
44. Sanchez RA, Ferris JP, Orgel LE (1967) J Mol Evol 30:223
45. Ferris JP, Orgel LE (1966) J Am Chem Soc 88:1074
46. Ferris JP, Kuder JE, Catalano AW (1969) Science 166:765
47. Ferris JP, Orgel LE (1966) J Am Chem Soc 88:1074
48. Eschenmoser A, Loewenthal E (1992) Chem Soc Rev 1
49. Sanchez R, Ferris JP, Orgel LE (1966) Science 153:72
50. Orgel L, Lohrmann R (1974) Acc Chem Res 7:368
51. Voet AB, Schwartz AW (1981) In: Wolman Y (ed) Origin of life. Reidel, Dordrecht, p 217
52. Miyakawa S, Cleaves HJ, Miller SL (2002) Origin Life Evol Biosphere 32:195
53. Proust JL (1807) J Chem Physik 384:1806
54. Ferris JP, Donner DB, Lotz W (1972) J Am Chem Soc 94:6968
55. Ferris JP, Donner DB, Lobo AP (1973) J Mol Biol 74:511
56. Ferris JP, Edelson EH, Auyeung JM, Joshi PC (1981) J Mol Evol 17:69
57. McKay RA Schaefer J, Steiskal EO, Ludicky R, Mattews CN (1984) Macromol 17:1124
58. Liebeman SA, Pesce-Rodriguez RA, Mattews CN (1994) Adv Space Res 15:71
59. Umemoto K, Takahashi M, Yokata K (1987) Origin Life Evol Biosphere 17:283
60. Völker TH (1960) Angew Chem 72:379
61. Ferris JP, Joshi PC, Edelson EH, Lawless JG (1978) J Mol Evol 11:293
62. Lieberman I, Kornberg A, Simms ES (1955) J Biol Chem 215:403
63. Grant NH, Alburn HE (1965) Science 150:1589
64. Orò J (1963) Nature 197:971
65. Bada JL, Lazcano A (2002) Science 1982
66. Coates J, Hartshorne NH (1931) J Chem Soc 657:193
67. Miller SL, Orgel LE (1974) In: The origin of life on the earth. Prentice-Hall, New York
68. Miyakawa S, Cleaves HJ, Miller SL (2002) Origin Life Evol Biosphere 32:195
69. Miyakawa S, Cleaves HJ, Miller SL (2002) Origin Life Evol Biosphere 32:209
70. Yamada H, Okamoto T (1972) Chem Pharm Bull 20:623
71. Yamada H, Okamoto T (1975) Yakugaku Zasshi 95:493
72. ECT Kirk-Othmer Encyclopedia of Chemical Technology (1978) Formic acid and derivatives. Wiley Interscience, Hoboken, NJ
73. Yamada H, Hirobe M, Higashiyama K, Takahashi H, Suzuki KT (1978) Tetrahedron Lett 42:4039

74. Yamada H, Hirobe M, Higashiyama K, Takahashi H, Suzuki H (1978) J Am Chem Soc 100
75. Zubay G (2001) Origin Life Evol Biosphere 31:87
76. Hill A, Orgel LE (2002) Origins Life Evol Biosphere 32:99
77. Shapiro R, Klein RS (1966) Biochem 5:2358
78. Sanchez R, Ferris JP, Orgel LE (1968) J Mol Biol 33:693
79. Sanchez R, Ferris JP, Orgel LE (1966) Science 154:784
80. Hirose Y, Ohmuro K, Saigoh M, Nakayama T, Yamagata Y (1991) Origins Life Evol Biosphere 20:471
81. Sanchez R, Ferris JP, Orgel LE (1968) J Mol Biol 33:693
82. Halmann M, Bloch S (1979) BioSyst 11:227
83. Robertson MP, Miller SL (1995) Nature 375:772
84. Ferris JP, Zamek OS, Altbuch AM, Freiman HJ (1974) J Mol Evol 3:301
85. Robertson PM, Levy M, Miller SL (1996) J Mol Evol 43:543
86. Schwartz AW, Chittenden GJF (1977) BioSyst 9:87
87. Sowerby SJ, Cohn CA, Heckl WM, Holm NG (2001) PNAS 98:820
88. Cohn CA, Hansson TK, Larsson HS, Sowerby SJ, Holm NG (2001) Astrobiol 1:477
89. Sowerby SJ, Petersen GB, Holm NG (2002) Origin Life Evol Biosphere 32:35
90. Saladino R, Crestini C, Costanzo G, Negri R, Di Mauro E (2001) Bioorg Med Chem 9:1249
91. Saladino R, Crestini C, Ciciriello F, Costanzo G, Negri R, Di Mauro E (2001) In: Astrobiology: future perspectives. Kluwer, Dordrecht, pp 393–413
92. Kirkpatrick E (1944) US Patent to EI Dupont de Nemours Co. Inc 2:346
93. Varma RS (2000) Tetrahedron 58:1235
94. Bernal JD (1951) In: The physical basis of life. Routledge Kegen Paul, London
95. Cairns-Smith C (1992) In: Tran Thanh Van J, Mounolou JC, Schneider J, McKay C (eds) Frontiers of life. Editions Frontieres, Gif-sur-Yvette, France, pp 119–132
96. Lorenz MG, Arderma AJ, Wackernagel WJ (1988) Genet Microbiol 134:107
Gallori E, Bazzicalupo M, Dal Canto L, Fani R, Nannipieri P, Vettori C, Stotzky G (1994) FEMS Microbiol Ecol 15:119
Pietramellara G, Franchi M, Gallori E, Nannipieri P (2001) Biol Fertil Soils 33:402
Franchi M, Bramanti E, Bonzi LM, Orioli PL, Vettori C, Gallori E (1999)Origin Life Evol Biosphere 29:297
Vettori C, Paffetti D, Pietramellara G, Stotzky G, Gallori E (1996) FEMS Microbiol Ecol 20:251
97. Saladino R, Crestini C, Ciambecchini U, Ciciriello F, Costanzo G, Di Mauro E (2004) ChemBiochem 8:1425
98. Negri R, Costanzo G, Saladino R, Di Mauro E (1996) Bio Techniques 21:910
Negri R, Ferraboli S, Barlati S, Di Mauro E (1994) Nucleic Acids Res 22:111
99. Saladino R, Minzione E, Crestini C, Negri R, Di Mauro E, G Costanzo (1996) J Am Chem Soc 118:5615
Saladino R, Crestini C, Minzione E, Costanzo G, Di Mauro E, Negri R (1997) Bioorg Med Chem 5:2041
100. Butlerow A (1861) Liebigs Ann Chem 120:295
101. Friesen DA, Haedely JV, Langford CH (1999) Environ Sci Technol 33:3193
102. Saladino R, Ciambecchini U, Crestini C, Costanzo G, Negri R, Di Mauro E (2003) ChemBiochem 4:514
103. Reis TLV (1965) Prog Nucleic Acid Res Mol Biol 4:189
104. Reid C, Orgel LE, Ponnampeuma C (1967) Nature 216:936
105. Fuller WD, Sanchez RA, Orgel LE (1972) J Mol Biol 67:25

106. Fuller WD, Sanchez RA, Orgel LE (1972) J Mol Evol 1:249
107. Kuzicheva EA, Tsupkina NV (1978) Zhurnal Evolyutsionnoi Biokhimii I Fiziologii 14 3:213
108. Deker P, Schweer H, Pohlman R (1982) J Chromatogr 244:281
109. Shapiro R (1988) Origin Life Evol Biosphere 18:71
110. Pitsch S, Eschenmoser A, Gedulin B, Hui S, Arrhenius G (1995) Origin Life Evol Biosphere 25:297
111. Zubay G (1998) Origin Life Evol Biosphere 28:13
112. Ricardo A, Carrigan MA, Olcott AN, Benner SA (2004) Science 303:196
113. Isbell HS, Frush HL, Wade CWR, Hunter CE (1969) Carbohydr Res 9:163
114. Dworkin JP, Miller SL (2000) Carbohydr Res 329:359
115. Springsteen G, Joyce GF (2004) J Am Chem Soc 126:9578
116. Sanchez RA, Orgel LE (1970) J Mol Biol 47:531
117. Tapiero CM, Nagyvary J (1971) Nature 231:42
118. Rabinowitz J, Sherwood C, Ponnamperuma C (1968) Nature 442
119. Lohrmann R, Orgel LE (1968) Science 161:64
120. Schoffstall AM (1976) Origin Life 7:399
121. Schoffstall AM, Kokko B (1978) In: Noda H (ed) Origin of life. Center for Academic Publications, Japan, p 193
122. Ponnamperuma C, Mack R (1965) Science 148:124
123. Schoffstall AM, Barto RJ, Ramos DI (1979) Origin Life 12:143
124. Becker B (1970) J Chem Eng Data 15:31
125. Philipp M, Seliger H (1977) Naturwissen 64:273
126. Knollmeller KO (1967) US Patent 3 33921; Chem Abstr 67, 74990c
127. Daub GG, Reid JD (1953) US Patent 2 609360; Chem Abstr 47, 920h
128. Lohrmann R, Orgel LE (1971) Science 171:490
129. Schwartz AW (1972) Biochem Biophys Acta 281:477
130. Verlander MS, Orgel LE (1974) J Mol Evol 3:115
131. Renz M, Lohrman R, Orgel LE (1971) Biochim Biophys Acta 240:463
132. Usher D, McHale AH (1976) Science 192:53
133. Uesugi S, Ikehara MN (1977) Biochem 16:493
134. Usher DA (1972) Nature New Biol 235:207
135. Usher DA, McHale AH (1976) Proc Nat Acad Sci USA 73:1149
136. Reiman R, Zubay G (1999) Origin Life Evol Biosphere 29:229
137. Schwartz AW (1969) Chem Commun 1393
138. Tsuhoko M, Fujimoto M, Ohashi S, Narai H, Motooka I (1984) Bull Chem Soc Jpn 57:3274
139. Tsuhoko M, Fujimoto M, Ohashi S (1981) Chem Lett 849
140. Etaix E, Orgel LE (1978) J Carbohydrates-Nucleosides-Nucleotides 5(2):91
141. Yamagata Y, Inoue H, Inomata K (1995) Origin Life Evol Biosphere 25:47
142. Cheng C, Fan CRW, Tong C, Miao Z, Chen J, Zhao Y (2002) Origin Life Evol Biosphere 32:219
143. Ponnamperuma C, Sagan C, Mariner R (1963) Nature 199:222
144. Steinman G, Kenyon DH, Calvin M (1965) Nature 206:707
145. Schwartz A, Ponnamperuma C (1968) Nature 218:443
146. Lohrmann R, Orgel LE (1971) Science 171:490
147. Handschuh GJ, Lohrmann R, Orgel LE (1973) J Mol Evol 2:251
148. Lohrmann R (1977) J Mol Evol 10:137
149. Gao K, Orgel LE (2000) Origin Life Evol Biosphere 30:45
150. Yamagata Y (1999) Origin Life Evol Biosphere 29:511

151. Ferris JP, Ryan TJ (1973) J Org Chem 38:3302
152. Ferris JP, Eldenson EH, Mount NM, Sullivan AE (1979) J Mol Evol 13:317
153. Ferris JP, Hagan WJ, Alwis KW, McCrea J (1982) J Mol Evol 18:304
154. Ferris JP, Yanagawa M, Hagan WJ (1984) Origin Life 14:99
155. Schramm G, Groetsch H, Pollmann W (1961) Angew Chem 73:610
156. Schramm G, Groetsch H, Pollmann W (1965) In: Fox S (ed) The origins of prebiological systems. Academic, New York, p 299
157. Kochetkov NK, Budowsky EI, Domkin VD, Kuromov-Borissov NN (1964) Biochim Biophys Acta 80:145
158. Jacob TM, Khorana HG (1964) J Am Chem Soc 86:1630
159. Gottikh BP, Slutsky I (1964) Biochim Biophys Acta 87:163
160. Fox SW, Harada K (1961) Science 133:1923
161. Sulston J, Lohrmann R, Orgel LE, Schneider-Bernloehr H, Weimann BJ, Miles HT (1969) J Mol Biol 40:227
162. Schneider-Bernloehr H, Lohrmann R, Sulston J, Weimann BJ, Orgel LE (1968) J Mol Biol 37:151
163. Orgel LE (1969) In: Sixth Fed Eur Biochem Soc Meeting, Madrid, Spain, p 37
164. Inoue T, Orgel LE (1971) Science 171:490
165. Osterberg R, Orgel LE, Sanchez RA (1973) J Mol Evol 2:231
166. Weimann BJ, Lohrmann R, Orgel LE, Schneider-Bernloehr H, Sulston JE (1968) Science 161:387
167. Lohrmann RH, Orgel LE (1978) Tetrahedron 34:853
168. Szostak JW, Ellington AD (1993) In: The RNA world. Cold Spring Harbour Laboratory Press, Cold Spring Harbour, NY, pp 511–533
169. Schwartz AW, Orgel LE (1985) Science 228:585
170. Van Vliet MJ, Visscher J, Schwartz AW (1995) J Mol Evol 41:261
171. Sawai H, Kuroda K, Hojo H (1989) Bull Chem Soc Jpn 62
172. Sievers D, Von Kiedrowski G (1994) Nature 369:221
173. Kanavarioti A (1994) Origin Life Evol Biosphere 24:479
174. Ferris JP, Yanagawa H, Dudgeon PA, Hagan JWJ, Mallare TE (1984) Origin Life 15:29
175. Ferris JP (1993) Origin Life Evol Biosphere 23:307—315
176. Ferris JP, Ertem G (1992) Science 257:1387
177. Ferris JP, Ertem G (1992) Origin Life Evol Biosphere 22:369
178. Ferris JP, Ertem G (1993) Origin Life Evol Biosphere 23:229
179. Ferris JP, Ertem G (1993) J Am Chem Soc 115:12270
180. Ertem G, Ferris JP (1996) Nature 379:238
181. Probahar KJ, Cole TD, Ferris JP (1994) J Am Chem Soc 116:10914
182. Probahar KJ, Ferris JP (1997) J Am Chem Soc 119:4330
183. Kawamura K, JP Ferris (1994) J Am Chem Soc 116:7564—7572
184. Probahar KJ, Cole TD, Ferris JP (1994) J Am Chem Soc 116:10914–10920
185. Kawamura K, Ferris JP (1996) Origins Life Evol Biosphere 26:151
186. Greares MP, Wilson WJ (1969) Soil Biol Biochem 1:317
187. Sewhney BL, Singh SS (1997) Clay Minerals 45:333
188. Kawamura K, Ferris JP (1999) Origins Life Evol Biosphere 29:63
189. Matsuno K (2000) In: Akaboshi M, Fujii N, Navarro-Gonzales R (eds) The role of radiation in the origin and evolution of life. Kyoto University Press, Kyoto, pp 9–24
190. Matsuno K (1997) Viva Origino 25:191–204
191. Ogasawara H, Yoshida A, Imai EI, Honda H, Hatori K, Matsuno K (2000) Origins Life Evol Biosphere 30:519
192. Imai E, Honda H, Hatori K, Matsuno K (1999) Origins Life Evol Biosphere 29:249

193. Kanavarioti A, Monnard PA, Deamer DW (2001) Astrobiol 1:271
194. Monnard PA, Kanavarioti A, Deamer DW (2003) J Am Chem Soc 125:13734
195. Sleeper HL, Orgel LE (1979) J Mol Evol 12:357
196. Sawai H, Ohno M (1981) Chem Pharm Bull 29:2237
197. Eschenmoser A (1997) Origins Life Evol Biosphere 27:535
198. Böhringer M, Roth HJ, Hunz J, Göbel M, Krishman R, Giger A, Schweizer B, Schreiber J, Leumann C, Eschenmoser A (1992) Helv Chim Acta 75:1416
199. Hunziker J, Roth HJ, Böhringer M, Giger A, Diederichsen U, Göbel M, Krishnan M, Jaun B, Leumann C, Eschenmoser A (1993) Helv Chim Acta 76:259
200. Otting G, Billeter M, Wuthrich K, Roth HJ, Leumann C, Eschenmoser A (1993) Helv Chim Acta 76:2701
201. Groebke K, Hunziker J, Fraser W, Peng L, Diederichsen U, Zimmermann K, Holzner A, Leumann C, Eschenmoser A (1998) Helv Chim Acta 81:375
202. Pitsch S, Wendeborn S, Jaun B, Eschenmoser A (1993) Helv Chim Acta 76:2161
203. Schlonvogt I, Pitsch S, Lesueur C, Eschenmoser A, Jaun B, Wolf RM (1996) Helv Chim Acta 79:2316
204. Reck F, Wippo H, Kudick R, Krishnamurthy R, Eschenmoser A (2001) Helv Chim Acta 84:1778
205. Wagner T, Huynh HK, Krishnamurthy R, Eschenmoser A (2002) Helv Chim Acta 85:4111
206. Beier M, Reck F, Wagner T, Krishnamurthy R, Eschenmoser A (1999) Science 283:699–703
207. Wippo H, Reck F, Kudick R, Ramaseshan M, Ceulemans G, Bolli M, Krishnamurthy R, Eschenmoser A (2001) Bioorg Med Chem 9:2411
208. Schöning KU, Scholz P, Guntha S, Wu X, Krishnamurthy R, Eschenmoser A (2000) Science 290:1347
209. Schöning KU, Scholz P, Guntha S, Wu X, Delgado G, Krishnamurthy R, Eschenmoser A (2002) Helv Chim Acta 85:4111–4153
210. Eschenmoser A (2004) Origins Life Evol Biosphere 34:277
211. Wu X, Delgado G, Krishnamurthy R, Eschenmoser A (2002) Org Lett 4:1283
212. Wu X, Guntha S, Ferencic M, Krishnamurthy R, Eschenmoser A (2002) Org Lett 4:1279
213. Kojima N, Bruce TC (2000) Org Lett 2:81
214. Nielsen PE (1999) Acc Chem Res 32:624
215. Han B, Wang Z, Krishnamurthy R, Juan B, Eschenmoser A (2000) Org Lett 4:3275
216. Kozlov IA, Politis PK, Pitsch S, Herdewijn P, Orgel LE (1999) J Am Chem Soc 121:1108
217. Kozlov IA, Zielinski M, Allart B, Kerremans L, Van Aerschot A, Busson R, Herdewijn P, Orgel LE (2000) Chem Eur J 6:151
218. Egholm M, Buchardt O, Nielsen PE, Berg RH (1992) J Am Chem Soc 114:1895
219. Micklefield J (2001) Curr Med Chem 8:1157
220. Nielsen PE, Egholm M, Berg RH, Buchardt O (1991) Science 254:1497
221. Orgel LE (1998) Trends Biochem Sci 23(12):491
222. Nelson KE, Levy M, Miller SL (2000) PNAS 97:3868
223. Schmidt JG, Nielsen PE, Orgel LE (1997) Nucleic Acids Res 25(23):4797
224. Böhler C, Nielsen PE, Orgel LE (1995) Nature 376(6541):578
225. Demidov VV, Yavnilovich MV, Belotserkovskii BP, Frank-Kamenetskii MD, Nielsen PE (1995) Proc Natl Acad Sci USA 92(7):2637
226. Koppelhus U, Nielsen PE (2003) Adv Drug Deliv Rev 55(2):267

227. Demidov VV, Broude NE, Lavrentieva-Smolina IV, Kuhn H, Frank-Kamenetskii MD (2001) Chembiochem 2(2):133
228. Protozanova E, Demidov VV, Soldatenkov V, Chasovskikh S, Frank-Kamenetskii MD (2002) EMBO Rep 3(10):956
229. Koppitz M, Nielsen PE (1998) J Am Chem Soc 120:4563
230. Izvolsky KI, Demidov VV, Nielsen PE, Frank-Kamenetskii MD (2000) Biochem 39(35):10908
231. Schneider KC, Benner SA (1990) J Am Chem Soc 112:453
232. Watson JD, Crick FH (1953) Nature 171:7370
233. Shapiro R, Kang S (1969) Biochem 8:1806
234. Garrett ER, Mehta PJ (1972) J Am Chem Soc 94(24):8532
235. Lindahl T, Karlstrom O (1973) Biochem 12(25):5151

Top Curr Chem (2005) 259: 69–122
DOI 10.1007/b136707

Published online: 11 August 2005

From the Prebiotic Synthesis of α-Amino Acids Towards a Primitive Translation Apparatus for the Synthesis of Peptides

Robert Pascal (✉) · Laurent Boiteau · Auguste Commeyras

Dynamique des Systèmes Biomoléculaires Complexes, UMR 5073 – CC 017, Département de Chimie, Université Montpellier 2, Place Eugène Bataillon, 34095 Montpellier, France
rpascal@univ-montp2.fr, laurent.boiteau@univ-montp2.fr, acommeyras@univ-montp2.fr

1 **Introduction** 70

2 **Formation of Amino Acids and Related Compounds** 73
2.1 Endogenous Formation (on the Primitive Earth) 73
2.1.1 Redox State on the Primitive Earth 74
2.1.2 The Strecker Reaction 75
2.1.3 Aldehyde Catalysis of α-Aminonitrile Hydrolysis 76
2.1.4 Bücherer–Bergs Reaction: Formation of Hydantoins and CAA 77
2.1.5 Amino Acid Formation in Submarine Hydrothermal Vents 78
2.1.6 Other Conditions and Pathways 79
2.2 Exogenous Formation and Delivery on Earth 79
2.2.1 Meteorites 79
2.2.2 Newly Identified Compounds in Meteorites 80
2.2.3 Micrometeorites 82
2.2.4 Amino Acids in the Interstellar Medium 82
2.2.5 Survivability of Exogenous Amino Acids 83

3 **Peptides** 84
3.1 Thermodynamic and Kinetic Stability of the Peptide Bond 84
3.2 Thermodynamically-Driven Peptide Formation from Unactivated Amino Acids 86
3.2.1 Salt-induced Peptide Formation 86
3.2.2 Dehydration in the Presence of Clays, Silica, or Alumina 87
3.2.3 Reactions of *N*-Phosphoryl Amino Acids 87
3.3 Polymerization of *N*-Carboxy Amino Acid Anhydrides (NCAs) 89
3.3.1 *N*-Carboxyanhydrides are Stabilized Compared to other Anhydrides 90
3.3.2 NCAs Are Kinetically as Reactive as Other Anhydrides 92
3.3.3 NCA Formation from Less Reactive Amino Acid Derivatives 93
3.3.4 NCA Formation from Reactive Amino Group Derivatives 95
3.3.5 NCA as the Most Activated Form of Amino Acids in an Aqueous Prebiotic Environment 95
3.3.6 NCAs as Intermediates in the Reactions of Much Less Activated Amino Acid Derivatives 96
3.3.7 New Pathways for NCA Prebiotic Synthesis 97
3.4 Polymerization of other Activated Amino Acids 99
3.4.1 Amino Acid-Phosphoric Acid Mixed Anhydrides 100
3.4.2 Amino Acid Thioesters and Amino Thioacids 104

3.4.3 Other Amino Acid Derivatives 105
3.5 Activating Agents 105
3.5.1 *N*, *N*′-Carbonyldiimidazole (CDI) 105
3.5.2 Other Activating Agents 106

4 Protometabolisms 107
4.1 Systems Based on Amino Acid and Peptides Only 107
4.2 Interactions between Amino Acids, Peptides, and Sugars 110
4.3 Did Biochemical Amino Acid Activation Evolve from an Early Nucleotide Activation Pathway? 110
4.4 Catalysis by Amino Acids and Peptides 114

5 Conclusion 114

Appendix 115

References 117

Abstract The demonstration that ribosomal peptide synthesis is a ribozyme-catalyzed reaction does not establish that amino acids and peptides have been useless and/or missing in an RNA world. The prebiotic chemistry of amino acids and peptides thus remains an important field of investigation. We first review the literature on amino acid synthesis and then focus on the topic of peptide synthesis pathways. Most proposed pathways are examined in the context of the thermodynamic constraints on peptide bond formation and using yet unpublished evaluations of the free energy contents of activated intermediates. This analysis provides evidences in favor of an important role of α-amino acid *N*-carboxyanhydrides (NCAs) as activated peptide monomers and possibly as early free energy carriers owing to their ability to activate inorganic phosphate and nucleotides. Finally, the transition towards the emergence of the biochemical amino acid activation pathways involving amino acid-phosphoric acid mixed anhydrides via both ribosomal and non-ribosomal processes is analyzed from the perspective that life emerged through a coevolution linking amino acid and nucleotide chemistries.

Keywords Amino acid synthesis · Peptide formation · Prebiotic chemistry · Coevolution

Abbreviations
AA α-amino acid
CAA *N*-carbamoyl α-amino acid (α-ureido acid)
CDI *N*,*N*′-carbonyldiimidazole
NCA amino acid-*N*-carboxyanhydride; *N*-carboxy-amino acid anhydride
PNA peptide nucleic acid

1 Introduction

Amino acids were the first prebiotic molecules to be identified more than 50 years ago in experiments reproducing hypothetic prebiotic environments [1–3] and they are probably the most easily synthesized. They can be formed in a variety of environments as shown by their identification in meteorites [4].

But, following the discovery of catalytic abilities of RNAs [8, 9], in the last twenty years much progress has been made in understanding the role of nucleic acid chemistry in prebiotic processes. RNAs have been shown to be capable of self-reproduction and of numerous catalytic activities. Then, it has been proposed [10–13] that modern living organisms using nucleic acid based information storage systems and enzymatic catalysis evolved from an earlier system called "RNA world" in which both functions would have been performed by RNAs (Fig. 1). The role of amino acids is even considered to have been quite unimportant by many supporters of this RNA world where living organisms used RNA for both catalysis and information storage [14]. How can nucleotides be formed and activated is a major objection to the fact that life began in an RNA world [14, 15]. In this context, the main questions to be answered are (i) what could have been the role of amino acids in a putative RNA world, or (ii) what could have been their role in the next stages of evolution that led to the modern DNA-protein world in which RNA is no longer the only information carrier?

The first hypothesis is that RNAs have used available amino acids to evolve from an RNA only world towards a nucleic acid-protein world. This hypothesis is in agreement with the role of RNA in the translation machinery, as for example the fact that the peptidyl transferase activity of the ribosome has been associated with the nucleic acid moiety and not the protein moiety [16, 17]. The driving force that guided the evolution from the RNA world towards the emergence of the translation machinery might have been that amino acids played a role of ribozyme cofactors [6, 7].

Another possibility is that the RNA world has been preceded by another information storage/catalytic function system (pre-RNA world, Fig. 1) that could for instance be derived from peptide analogues of nucleic acids called Peptide Nucleic Acids (PNAs) [15, 18–20]. However, there is no remaining evidence of a switch from this early information storage system into that based on nucleic acids.

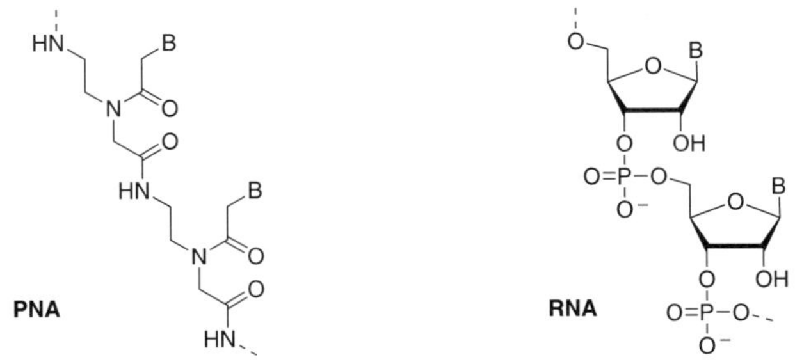

Scheme 1

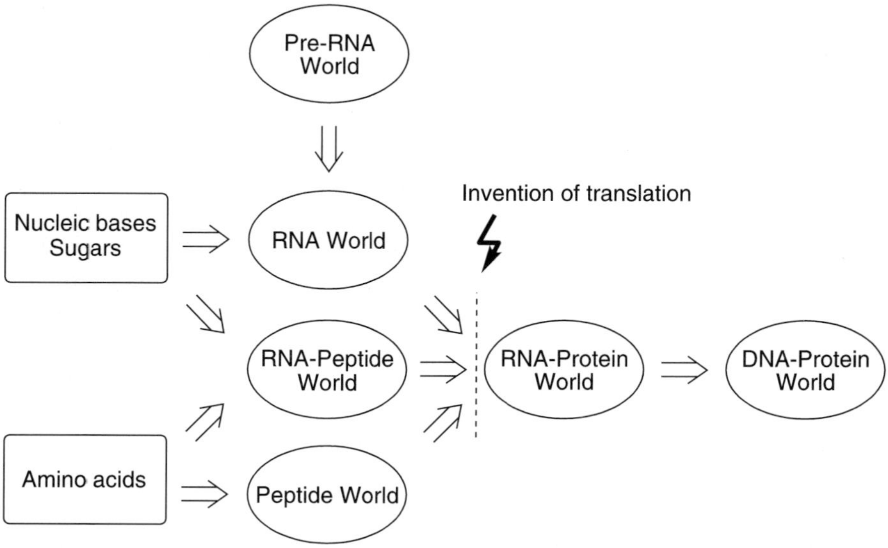

Fig. 1 The sequence of different hypothetical stages in the evolution from an abiotic chemical world towards the modern DNA-protein world. The transition from a world in which RNA was the carrier of genetic information (RNA-protein world) to the present DNA-based system is strongly supported by the fact that the only pathway of synthesis of desoxyribonucleotides found in living organisms starts from the corresponding ribonucleotides [5]. The previous stages considered here (RNA world, peptide world ...) are hypothetically deduced by simplifying the RNA-protein world. The hypothesis of an RNA world is supported by RNA catalytic abilities, but assistance by amino acids or peptides, for instance as ribozyme cofactors [6, 7], may have been helpful at that stage. The direct emergence of the genetic code from an RNA-free peptide world will not be considered further

There is a last possibility to circumvent the above-mentioned difficulties; it is that life arose from an early symbiosis between amino acid/peptide and nucleotide/RNA chemistries [21–24]. It is consistent with the fact that the translation machinery (ribosome, *t*RNAs) seems to be one of the oldest components of the cell as a result of the universality of the genetic code. The translation apparatus would then be considered as a remnant of this interaction. Analysis of biochemical pathways have provided indications in favor of a model based on an early ribonucleopeptide world involving the cooperation of oligopeptides and nucleotides in covalently bound conjugates [25, 26].

In this review we will consider the different aspects of the prebiotic chemistry of amino acid and peptides. It includes a survey of recent advances in amino acid formation pathways, the question of peptide bond formation considered both from the synthetic point of view and by taking into account the thermodynamic constraints on peptide bond formation and the availability of activated forms of amino acids and/or activating agents. The question of the emergence of the genetic code will be considered in the context of a coevolution process involving amino acid and nucleotide chemistries or at least a linkage between both systems. Since the metabolism of living organisms is based on energy transfers through chemically coupled reactions it is likely that this kind of processes has played a role in the early stages of chemical evolution in which a kind of protometamolism involved probably a sequence of spontaneous reactions having favorable free energies and favorable rate constants [27]. Our hypothesis is inspired from the study of non-linear chemical systems under far-from-equilibrium conditions [28] that can bring about some kind of organization when fed with chemical energy and matter and thus maintained in states that may be capable of evolving.

2 Formation of Amino Acids and Related Compounds

The availability of building materials (including amino acids) under hypothesized primitive Earth conditions is a major concern in prebiotic chemistry. Major – closely linked to each other – concerns in the field are "which compounds can be formed", "how much", and "through which mechanisms". When estimating the availability of prebiotic organics, two relevant, parallel pathways are usually pointed out:

- endogenous (in-situ) synthesis from simple molecular compounds, associated with energy sources such as impacts, radiation, lightning, heat ...
- exogenous formation in the interstellar medium (gas clouds, interplanetary dust particles, ices...), then delivery on Earth by e.g. comets, meteorites, micrometeorites, etc.

Both pathways probably involved quite different conditions, the main difference being the absence of liquid water in the interstellar medium. Nevertheless the basic building blocks and chemical reactions should have been roughly similar, thus leading to important connections between these two routes. While a wide variety of amino acids are prebiotically relevant (as attested by either Urey–Miller experiments or meteorite analysis), we shall focus in this section on α-amino acids (as the most relevant to biochemistry) and closely related compounds.

2.1 Endogenous Formation (on the Primitive Earth)

Very little is certain on organic matter formation in the atmosphere or in the ocean of the primitive Earth as a result of lightning, volcanism, or other energy sources. Since no preserved geological records exist from the concerned era, there is no certainty about the "reactor" conditions (physical parameters; kind, concentration and/or fluxes of reagents). The endogenous synthesis of amino acids entirely relies on models (built using knowledge from domains such as astronomy or geology), which must necessarily cover a wide variety of conditions.

Geological records (with the presence of hydrated minerals in the continental crust) lead to the conclusion that both large amounts of liquid water and emerged lands were present quite early after the formation of the Earth [29]. But prebiotic syntheses of complex compounds are unlikely to have occurred in the ocean itself, where the dilution factor heavily impedes further possibilities of intermolecular reactions, but rather in smaller lakes or in ponds where concentration processes could occur. Another investigated condition is the chemistry of species adsorbed onto mineral surface, where a pre-association can also occur. The main investigated conditions are gas-phase in presence of liquid water, aqueous phase, hydrothermal vents, and drying lagoons. Activation sources are mostly photochemistry (in the atmosphere), electric discharges, or heat associated with quenching (allowing the formation of products that are thermodynamically unstable at low temperature but become favored as temperature increases).

2.1.1 Redox State on the Primitive Earth

A main change in the last 20 years is related to the redox state of the primitive Earth atmosphere. Since Oparin, it was initially believed that, as the global solar system composition, the primitive Earth atmosphere was reducing (rich in e.g. hydrogen, methane, ammonia). This belief led to Miller to perform the successful experiment that demonstrated the possibility of amino acid production under conditions reproducing the primitive Earth [1]. Although the reduced model keeps some popularity, more recent models of the solar system and either geological, moon or meteorite records, argue in favor of a more neutral atmosphere, mostly made of N_2 and CO_2 [30]. This redox state is of high importance for prebiotic chemistry since a much richer chemistry (both in abundance and diversity of compounds) is observed in reducing media. In a reference review [31], Chyba and Sagan estimated that depending on the redox state of the atmosphere the endogenous production could vary by several orders of magnitude, so that the steady-state concentration of organic matter in the primitive ocean could vary from $0.4\,g\,L^{-1}$ (reducing atmosphere) to $0.4 \times 10^{-3}\,g\,L^{-1}$ (non-reducing atmosphere). Nevertheless the redox state is not a homogeneous property.[1] While the core of telluric planets – mostly made of metals – remains reducing, the upper atmo-

[1] At scales much larger than molecular where the electrical neutrality must be preserved, the redox state of a medium is not a matter of exceeding/missing electrons, but rather of electropositive/electronegative element ratio: an oxidized medium is rich in electronegative elements whereas a reduced medium is rich in electropositive elements (all metals, which represent the large majority of the periodic table).

sphere undergoes a continuous oxidation due to photochemistry (molecule dissociation with release of H$^\bullet$ radicals) combined with H$^\bullet$ escape towards outer space [32], while heavier oxidized moieties are retained in the atmosphere by gravity. Electric discharges (lightning in e.g. volcanic plumes) can also participate to such a local redox state change. Therefore even with a non-reducing atmosphere some local environments may remain favorable to prebiotic chemistry, such as volcanic plumes or submarine hydrothermal vents for instance, both releasing reduced compounds.

2.1.2
The Strecker Reaction

Since HCN and aldehydes were produced directly from the electric discharge in the Miller's experiment [33], the Strecker reaction was very early proposed as a likely pathway for the prebiotic synthesis of amino acids. This reaction discovered in 1850 [34] is the most anciently known abiotic synthesis of α-amino acids, it originally consisted in the formation of an α-aminonitrile **1** from a carbonyl compound (either aldehyde or ketone), ammonia and hydrogen cyanide in moderately alkaline aqueous solution followed by aminonitrile hydrolysis in strong acid.

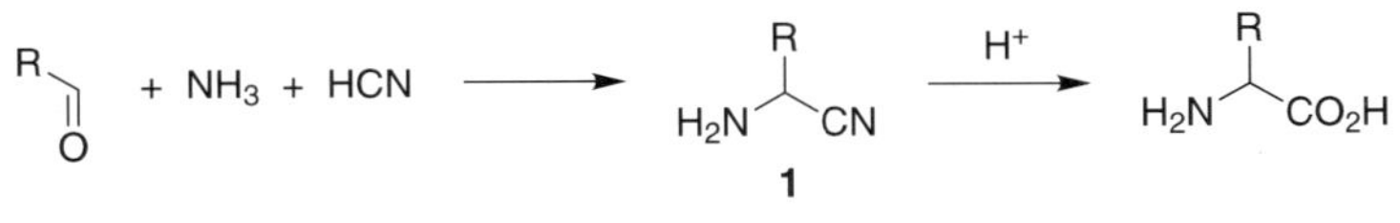

Scheme 2

The mixture of a carbonyl compound, ammonia and hydrogen cyanide equilibrates into aminonitrile **1** and cyanohydrine **2**, the product ratio being pH-dependent.

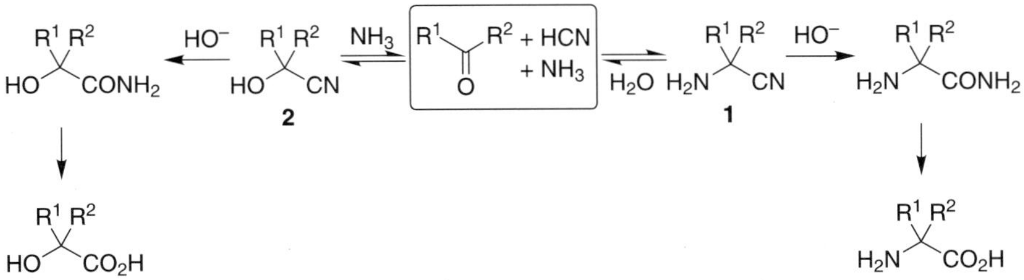

Scheme 3

Kinetic and thermodynamic studies have been carried out to determine the parameters of these systems [35–40]. These studies of the carbonyl compound-ammonia-HCN equilibrium also demonstrate the difference in reactivity between aldehydes and ketones [39]. The slow hydrolysis of the nitrile group leads to the formation of a mixture containing α-hydoxy- and α-amino acids. The hydroxy acid/amino acid ratio has been considered to reflect the cyanohydrine/aminonitrile equilibrium ratio, the nitrile hydrolysis rates being initially supposed to be rather similar for cyanohydrines and aminonitriles [41].

When the aldehyde content becomes significant, the system is complicated by a further reaction giving iminodinitriles **3** and iminodiacids **4** as subsequent hydrolysis products.

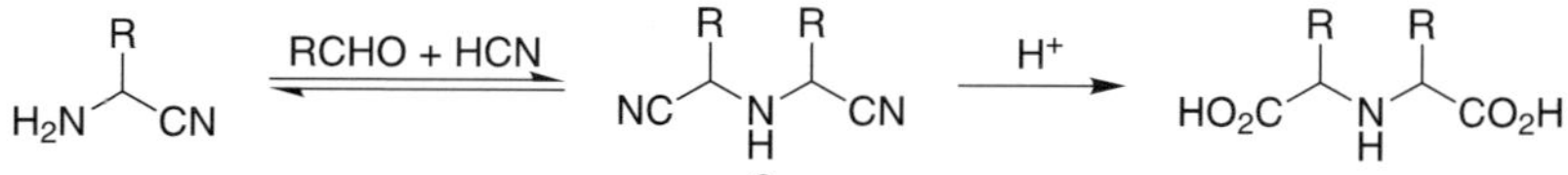

3
4

Scheme 4

However, these reactions may be basically seen as a core of equilibrated reactions between central components, cyanohydrins and aminonitriles, surrounded by non-equilibrated reactions. Although easily accessible to experiment, the core of reversible reactions is possibly not representative of the behavior of the whole system, especially concerning the final product repartition. Indeed such systems probably worked out of equilibrium with a continuous flow of building blocks. A kinetic selectivity is indeed observed owing to specific transformations of aminonitriles that are developed subsequently because of their importance in prebiotic chemistry.

2.1.3 Aldehyde Catalysis of α-Aminonitrile Hydrolysis

Kinetic and mechanistic studies done since the late 1970s by Commeyras and co-workers (with industrial development purposes in the amino acid field), provided new insights into Strecker-related reactions in the prebiotic field of interest. The story began with the discovery of a catalytic pathway involving carbonyl compounds for nitrile hydration into amide, which is specific of aminonitriles [42]. Aldehydes are better catalysts than ketones, formaldehyde being the most efficient, which is highly significant since it is a very likely prebiotic molecule. With a rate constant several orders of magnitude higher than the non-catalyzed pathway, this process yields AA even at low concentrations of ammonia [43], as it can be the case in mildly or non-reducing media. It also gives an answer to the Miller's observation that it was difficult to account for unexpected nitrile hydrolysis rates in his apparatus [33].

In the special case of a low, continuous flow of ammonia, this process allows the formation of substantial amounts of AA by displacement of the cyanohydrin-aminonitrile equilibrium. The cyanohydrin form can be consid-

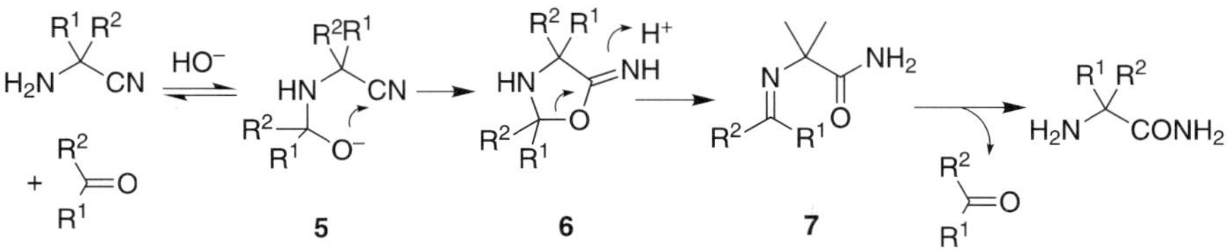

Scheme 5

ered as a storage of reactive compounds (aldehydes, ketones, HCN) under a much more stable form, thus allowing it to wait for ammonia supply.

The considerable increase in rate constant associated with this process has been attributed to the induced intramolecular attack in adduct **5** compared to the intermolecular attack of hydroxide ion that is entropically unfavorable [42]. The cyclic intermediate **6** readily rearranges into open-chain imine **7**. The similarity between this kind of mechanism and enzymatic catalysis has been associated with the use of binding energy to compensate for the entropy cost while coupling the carbonyl compound to the amino group [44] in a way that is similar to the utilization of favorable interaction with non-reacting portion of the substrates by enzymes [45].

2.1.4
Bücherer–Bergs Reaction: Formation of Hydantoins and CAA

A reaction discovered earlier but having very similar features is the Bücherer–Bergs reaction, which involves bicarbonate/CO_2 instead of a carbonyl compound and leads to hydantoin **11**. This reaction was originally developed from the 1930s [46] for applicative purposes and had not been considered as prebiotically relevant until recently [43]. This pathway involves the addition of CO_2 on the amino group [47], then the intramolecular attack of the nitrile group by the carbamate of adduct **8**. The resulting cyclic intermediate **9** undergoes a rearrangement into the stable hydantoin product **11** trough transient isocyanate **10** [48]. It is worth mentioning that the rearrangement requires the abstraction of a proton at N^3 in oxazolidine **9** so that the reaction is only observed from α-aminonitriles with a free amino group, yielding hydantoins without substituent at position 1.

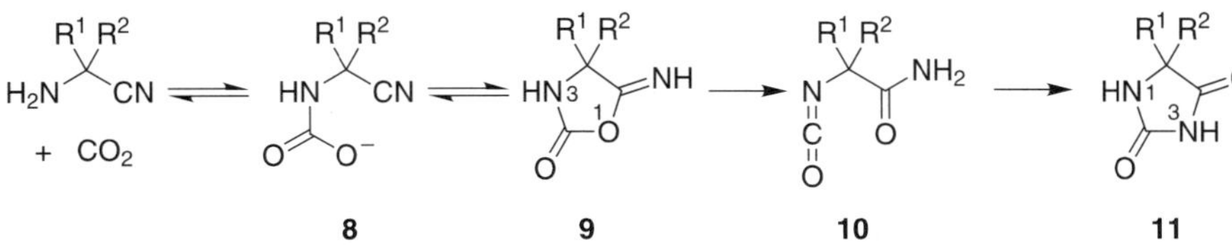

Scheme 6

CO_2 promotes the reaction of unreactive nitrile groups at moderate pH but cannot be truly considered as a catalyst since it is not released at the end of the reaction (unless the hydantoin undergoes further hydrolysis) unlike carbonyl compounds in α-aminonitrile hydration (Sect. 2.1.3). However, hydantoins can be considered as AA precursors through a two-step hydrolysis. The first step leads to an *N*-carbamoyl amino acid (CAA). The rate of this HO^--catalyzed step has been examined by Taillades and co-workers [43], extending the earlier work of Blagoeva et al. [49].

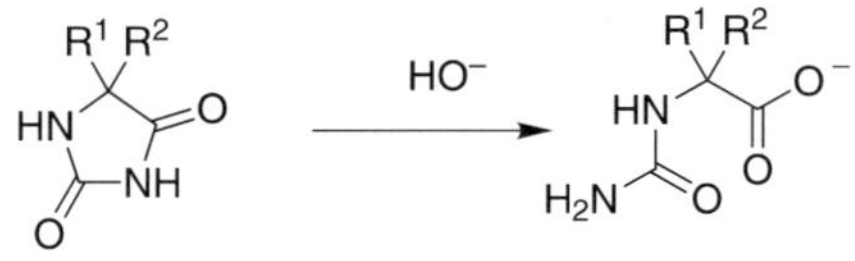

11 **CAA**

Scheme 7

Since an important feature of Bücherer–Bergs hydantoin formation is that the process can only work for α-aminonitriles without substituent on the amino group, it follows that one compound of the equilibrium mixture formed from an aldehyde, ammonia, and cyanide is selectively reacted through an irreversible process leaving *N*-alkylated aminonitriles or iminodinitriles unreacted. However, the difficulty with this process is that CAAs and hydantoins are poorly reactive towards hydrolysis and need long periods of time to be converted into free AAs. But, CAAs may also have per se a prebiotic importance in activation pathways towards polypeptides (see Sect. 3.3.7). CAAs can also be synthesized by reaction of free amino acids with cyanic acid/cyanate (a likely prebiotic compound [50]). In the presence of a steady-state concentration of either cyanate or urea in aqueous medium, CAAs are at equilibrium with AA [51].

2.1.5
Amino Acid Formation in Submarine Hydrothermal Vents

These environments provide favorable conditions with the following characteristics: high temperatures (100–350 °C), high pressure, reducing, and metal-rich environment. Under such conditions, Strecker, Bücherer–Bergs and related reactions are very likely to occur in most cases. However the chemical pathways are not very well characterized under such conditions and other pathways are also possible. Laboratory conditions simulating such environments have been investigated for more than 10 years [52, 53], already showing the formation of amino acids. However at higher temperatures (above 300 °C) amino acids turn out to be unstable, so that formation and degradation processes are competing [54]. Kobayashi and co-workers observed that quenching hot, supercritical aqueous fluids with cold sea water (as naturally occurring in undersea vents) can nevertheless lower AA degradation rates [55, 56].

2.1.6 Other Conditions and Pathways

After the reducing state of the primitive Earth atmosphere had been reconsidered, amino acid synthesis in non-reducing environments has been subject to investigation. Kobayashi and co-workers studied the activation of non-reducing gas mixtures (e.g. of N_2, CO_2, CO, in the presence of liquid water) by either high-energy particle bombardment (simulating the cosmic rays), UV or γ-ray irradiation [57–59], while Rode and co-workers examined the activation of similar mixtures by electric discharges [60]. Both of them observed amino acid formation.

Intermediates of the formose reaction can yield amino acids in the presence of ammonia and thiols [61, 62] through a process that will be examined in more details in subsequent sections (see Sect. 3.4.2) since amino-thioesters are activated peptide precursors.

2.2 Exogenous Formation and Delivery on Earth

Except recent radioastronomic observations, exogenous amino acids have for the moment only been known through the analysis of meteorite content, especially of the Murchison meteorite (the most organic-rich meteorite known to have fallen on Earth).

Plausible synthetic conditions of exogenous AA mainly comprise gas phase, grain-surface, bulk ice (e.g. comets) and planetary bodies. Except where condensed aqueous phases could occur (where Strecker-related reactions could take place), the chemistry involved is probably mostly photochemical and radical, using free atoms as building blocks, through non-selective, high-energy photo-activation and radical recombination of small molecular fragments. Such non-selectivity is supported by the variety of compounds identified in meteorites (see next section).

2.2.1 Meteorites

Amino acids – and more generally organic substances – are mainly detected in carbonaceous chondrites, a minor, carbon-rich class of meteorites. These are believed to have originated from parent bodies having underwent alteration by liquid water at some stages of their existence, as attested by geochemical studies [63].

The analysis of meteorite samples is an intricate problem because of the complexity of the matrix, of the low abundance (0–100 ppm) of searched organic compounds, and therefore of the possibility of terrestrial contamination (especially concerning biogenic compounds) during extraction/analysis process. The extraterrestrial origin is generally confirmed by isotopic analyses, namely of the $^{13}C/^{12}C$ and D/H ratios, those of authentic meteoritic samples being quite distinct from those of biogenic origin. Amino acids are generally detected after aqueous extraction. It has often been observed that detected amounts are higher after acidic extraction (e.g. heating above 100 °C for several hours in 6 N HCl), leading to the conclusion that not only amino acids, but at least precursors of them are present in meteorites, these precursors possibly being either hydantoins, aminonitriles, amidines or peptides.

The catalogue of organic substances detected in meteorites has been listed in detail a few years ago in several reference reviews [63, 64]. Subsequent extensions in the field of amino acids and related compounds are due to the progress in analytical technique sensitivity in the recent years. These – however minor – extensions allowed to identify some new, previously undetected compounds (see Sect. 2.2.2). The distribution of molecular structures (decreasing abundance with the number of carbon atoms in homologous series, and predominance of branched isomers) strongly argues for the involvement of non-selective processes (probably photochemical and/or radical) in the early steps of their formation. The amino acid set comprises more than 70 species measured at the 10–100 ppm (w/w) level [4], most of them being α-amino acids (including at least 8 proteogenic: Gly, Ala, Val, Leu, Ile, Pro, Asp, Glu, all detected as racemic). Since the analysis of the mineral matrices strongly suggests that they have been in contact with liquid water at some stage of their existence, the involvement of Strecker-related reactions in meteoritic amino acid formation is plausible.

2.2.2
Newly Identified Compounds in Meteorites

Peptides and Other Carboxamide Derivatives

Amides, *N*-alkyl amides, lactams, hydantoins have been identified in both Murchison and Yamoto-791198 meteorites [63, 65], confirming preliminary results from Cooper and Cronin [66]. The only peptides identified until now are diglycine and diketopiperazine (cyclic diglycine). Since only N^1-unsubstituted hydantoins have been detected, this provides a good argument for the involvement of the Bücherer–Bergs reaction in their formation (cf. Sect. 2.1.4).

Diamino Acids

Several of them have been recently identified by Meierhenrich et al. in the Murchison meteorite, at the 5–50 ppb level [67]. This discovery extends the meteoritic amino acid suite to alkaline ones, of which the proteogenic representatives, namely Arg and Lys, remain both undetected in meteorites however.

Imino Diacids

4 have been recently detected in the Murchison meteorite [68]. In association with α-hydroxy- and α-amino acids they provide a complete signature of the Strecker reaction as already observed in Miller's early experiments [33], thus giving an additional argument to its involvement in meteoritic α-amino acid synthesis.

Non-racemic Amino Acids

The enantiomeric excess of several α-alkyl amino acids from the Murchison meteorite has been measured by Cronin and Pizzarello [69–71], who found significant values (1–15% ee L). Conversely to previous estimates often biased by terrestrial contamination, these results can be considered as reliable since target AA are both non-biogenic and non-terrestrial in origin (the latter confirmed by D/H and $^{13}C/^{12}C$ isotopic ratios). The quaternary α-carbon prevents the racemization of the compounds, thus allowing the long-term preservation of the enantiomeric excess.

Tentative explanations for the origin of such ee include enantioselective photodegradation induced by ultra-violet circularly polarized light (UVCPL), or the parity-violation energy difference (PVED) that would stabilize very slightly the L enantiomer of amino acids. Neither has received a general agreement however. Moreover, although Strecker-related processes (involving ketones instead of aldehydes as the starting material) are quite plausible formation pathways, several parameters such as isotopic enrichment or hydoxy/amino acid abundance ratio show meteoritic α-alkyl AA as a distinct subset – thus non representative – from α-H amino acids [4].

The fact that the ee of meteoritic α-alkyl amino acids has the same (L) orientation as that of biogenic material has raised considerable investigation for (i) understanding the origin of such ee and (ii) designing amplification processes to make the transition between quite modest ees (a few percent) and biomolecular homochirality. As non-racemizable compounds, non racemic α-alkyl amino acids may have played a seeding role in further amplification processes [72, 73] (see Sect. 4.2).

2.2.3 Micrometeorites

These dust-grain-sized bodies (all samples being collected in Antarctic ices where they are better preserved, and easier to collect and identify) may have been of significant contribution for organic matter/amino acid delivery on the primitive Earth. Compared to meteorites, higher amounts of them fall on the Earth (ca. 20 000 tons per year); a higher fraction of them (ca. 80% of identified micrometeorites) belong to the carbonaceous chondrite family, and exhibit a higher amino acid content than the Murchison meteorite [74, 75]. Moreover the samples collected showed minor to inexistent alteration related to atmosphere entry (no heating). Recent investigations led to the conclusion that micrometeorites are very probably a distinct (possibly more ancient) class of material, compared to chondrites [76].

2.2.4 Amino Acids in the Interstellar Medium

Works initiated several decades ago on the detection of molecular species through microwave-range spectroscopy [77], recently succeeded with the detection of glycine in interstellar dense molecular clouds [78–80]. The understanding of interstellar medium chemistry mostly relies on theoretical (quantum) chemistry calculations, to predict either possible reactions under such interstellar conditions, or the spectra of thus-produced chemical species. It is commonly considered that synthesis is unlikely to occur directly in the gas phase, but rather in the solid state (onto or inside solid ice/grain) [81, 82]. Nevertheless, Kobayashi and co-workers succeeded in synthesizing amino acid precursors (amino acids being observed after acidic hydrolysis) in the gas phase (80 °C, near atmospheric pressure), by either UV or γ-ray irradiation of mixtures of water, methanol and either nitrogen or ammonia; such conditions being however quite far from those of interstellar medium [83, 84].

An important feature of most interstellar medium conditions is that water (whenever present) does not exist as a liquid. This is for instance the case of comets, the chemistry of which is probably very rich although yet poorly known except through laboratory investigation of experimental models. These – so-called interstellar ice analogues – are usually mixtures of fundamental molecular species such as H_2O, CO, CO_2, NH_3, N_2, MeOH ..., often vacuum-deposited and studied in the 10–100 K range. When irradiated with UV [84–86], γ-ray [84] or bombarded by high-energy protons (the latter simulating cosmic rays) [84, 87, 88], the formation of various amino acids (most of those found in meteorites) or their precursors was observed in most cases[2] [84, 85, 87–89]. Unsurprisingly, higher-energy activation appeared necessary when N_2 was the nitrogen source instead of NH_3 [89]. Partial asymmetric induction was observed when irradiating with UVCPL [86].

Apart from ice chemistry, Devienne and co-workers observed the formation of a series of amino acids (Gly, Ala, Leu, Ile, Asn, Asp, Glu, His, Phe, Arg, Tyr, and either Lys or Gln) and of the dipeptide Gly-Ala, as minor products among various organic compounds after bombardment of carbon targets by high-energy molecular jets of H_2/N_2 in presence of oxygen. After sparkling the target with sulfur atoms, S-containing amino acids (Cys, Met, homo-Cys and S-methyl Cys) were also observed [90, 91]. These results suggest that interstellar (in either grains or molecular clouds) organic chemistry may be much richer than currently believed.

2.2.5
Survivability of Exogenous Amino Acids

The survivability of AA over long periods under interstellar medium conditions (where exposed to hard radiations), or during atmospheric entry (where heated) has been subject to a few studies. Amino acid photostability has been examined through both laboratory [92, 93] and in-situ experiments aboard satellites or space stations [94–96]. Both lead to the conclusion that when exposed to UV most AA (as well as oligopeptides, with exception of Gly and diketopiperazine) are photolabile on the long term, unless shielded from radiations by an opaque mineral matrix (i.e. being inside grains). The thermolability of AA in the solid state has been examined by Navarro-Gonzalez and co-workers [97], through rapid heating in N_2/CO_2 atmospheres, to simulate atmospheric entry and/or meteoritic impact conditions. While no organic compound survives temperatures above 700 °C, AA are still recovered at the few-percent level after heating to the 500–600 °C range. The same authors observed that in the presence of silicate minerals, amino acid pyrolysis affords (through condensation-dehydration reactions) highly thermostable cyclic amidines such as **12** and **13** [97, 98]. It is worth noting that their acidic hydrolysis (by e.g. 6 N HCl) reverts the free amino acids.

[2] As for meteorite analysis, amino acids from such interstellar ice analogues are usually detected after (often acidic) hydrolysis of crude reaction products, and subject to the same interpretation restrictions. Pointing out the formation of peptides [85] (without further arguments) to explain the detection of amino acids in such acid-hydrolyzed irradiated ices is at least questionable since many other precursors can be involved.

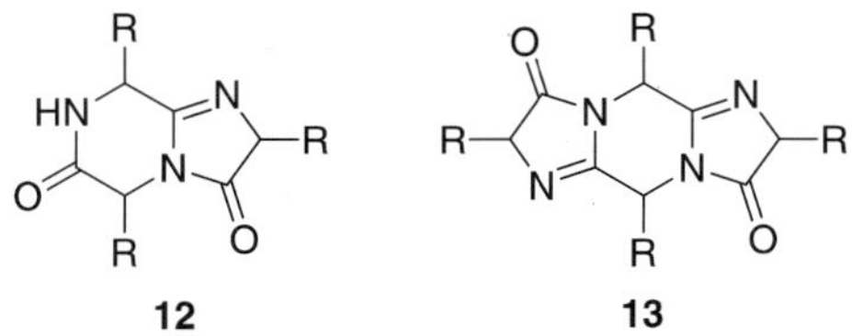

Scheme 8

3 Peptides

Before presenting recent results on peptide bond formation under prebiotic conditions it is useful to consider the kinetic and thermodynamic stabilities of peptide bonds that correspond to the physico-chemical constraints to the survival of peptides. Indeed, prebiotic peptides are likely to have been exposed to the primitive environment during long periods of time. However, it should be kept in mind that life being based on a continuous process of generation and breakdown of biomolecules, any primitive living organism would have involved disintegration pathways. Then the physico-chemical endurance of biomolecules must be considered from a dynamic perspective considering a steady state favorable to the emergence and the evolution of new properties rather than from a static perspective.

3.1 Thermodynamic and Kinetic Stability of the Peptide Bond

There is a general belief that peptides are not thermodynamically stable [99], but the formation of an internal bond in a peptide chain actually corresponds to a process that is not far from equilibrium ($K_{pep} \sim 0.1\ M^{-1}$) and enzymatic peptide synthesis can yield peptides in satisfactory yield provided that the product is removed from the solution by precipitation or other systems of phase partition [100, 101].

Scheme 9

This means that peptide bond formation is not a totally unfavorable process but actually corresponds to a change in standard free energy that is close to 0 [102]. But this assertion is true within a rather limited pH range only (Fig. 2), where peptide bond formation is a very slow process in the absence of catalyst so that the reversibility of peptide bond hydrolysis revealed by enzymatic catalysis is usually not accessible with purely chemical systems.

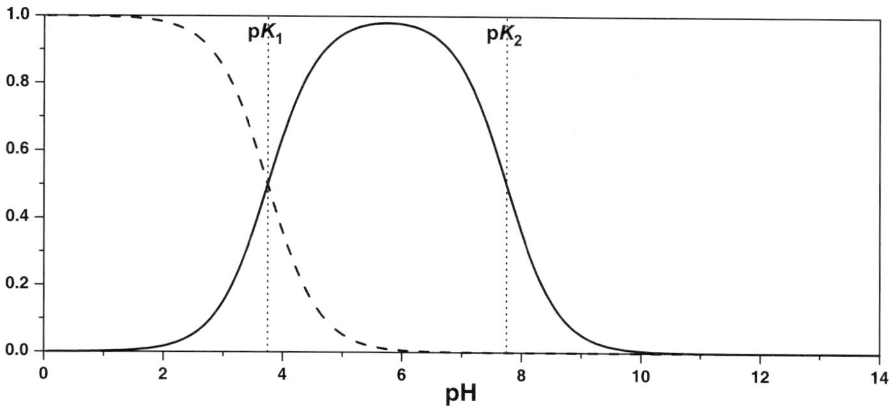

Fig. 2 Theoretical influence of pH on the ratio of equilibrium constants K_{app}/K_{pep} and K_{app}/K_{est} for peptide (*solid line*) and ester (*dotted line*) bond formation, respectively. pK_1 and pK_2 are the acid dissociation constants of the carboxyl and amino groups, respectively, the values used for the calculation (pK_1 = 3.75, pK_2 = 7.75) are representative of an internal peptide bond. Theoretical values calculated using the following equations: $K_{app}/K_{pep} = 1/(1 + [H^+]/K_1)(1 + K_2/[H^+])$ with $K_{pep} = [\text{peptide}]/[RCO_2^-][R'NH_3^+]$, $K_{app}/K_{est} = 1/(1 + K_1/[H^+])$ with $K_{est} = [\text{ester}]/[RCO_2H][R'OH]$

This behavior is different from that of usual aliphatic esters (Fig. 2), which are more stable at low pH and, therefore, which can generally be synthesized in strong acids without activating agent. This is because both components of the bimolecular reaction are present under their reactive forms (neutral carboxylic acid and alcohol) at these pH values, which are favorable to high rates because of acid catalysis. This situation is never found over the whole pH range for peptides because the acidities of the *C*-terminal carboxyl group (p$K_A \sim 3.5$–4) and of the *N*-terminal amine (p$K_A \sim 7.5$–8) to be ligated do not match. The near stability of peptide bonds in the pH range 4–7.5 seems to be unexpected since peptide bonds require in some instances strong reagents to be formed in high yields. Actually, the difficulty in forming peptide bonds is rather associated to a kinetic barrier than to a thermodynamic barrier. In the reverse direction, the kinetic stability towards hydrolysis is clearly an advantage from the biological point of view and the usual lifetime of a peptide bond has been assessed to be of the order of hundreds of years in neutral water and at usual temperatures [103, 104]. It is also illustrated by the difficulty in the selection of artificial catalysts capable of hydrolyzing the peptide bond [105] at moderate pH and that could be used to replace or imitate proteolytic enzymes.

From the point of view of prebiotic chemistry, it can be concluded from the thermodynamic and kinetic stability of amides that the formation of prebiotic peptides has very probably required catalysts and/or dehydrating or activating agents. But their spontaneous formation could not be completely excluded under certain conditions though, in the absence of catalyst, it would have required periods of the same order of magnitude as their lifetime (10^2–10^3 years).

3.2 Thermodynamically-Driven Peptide Formation from Unactivated Amino Acids

Since peptide bonds are not highly unstable from a thermodynamical point of view, their stability can be increased by subtle changes in the physicochemical properties of the synthesis medium, which can have an influence on the equilibrium constant or on the activity of reactants and of water. Fluctuating environment with heating and the presence of minerals have indeed been shown to induce peptide bond formation for a long time [106]. This is consistent with the fact that elevated temperatures seem to be thermodynamically favorable to peptide bond formation [107]. But systems that are efficient at moderate temperature require the presence of a catalyst. Two systems involving original means of catalysis have been developed in the last decade. One is based on metallic ion catalysis and the other one on the presence of a phosphoryl neighboring group lowering the transition state of peptide bond formation (the backward reaction of peptide bond hydrolysis). Clay catalysis is still a field of investigation.

3.2.1 Salt-induced Peptide Formation

Salt induced peptide formation [108] is based on a "dehydrating activity" of concentrated NaCl solutions in which free water molecules are less available, which drives the equilibrium towards peptide formation. This change in the thermodynamic barrier is coupled to a decrease of the kinetic barrier owing to the addition of a Cu(II) salt catalyst. A complex of copper with two amino acid ligands **14** has been proposed to be responsible for the catalytic process. In this way, the reaction between two amino acid ligands leading to peptide bond formation can take place intramolecularly:

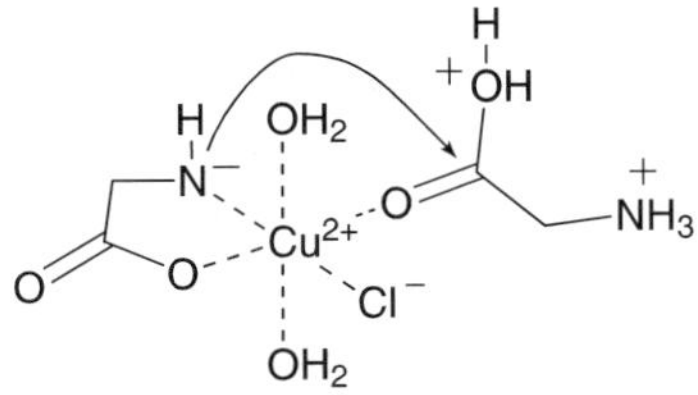

14

Scheme 10

A massive increase in rate [109] can be expected from this intramolecular reaction. The presence of glycine and H-Gly-Gly-OH has been shown to increase the incorporation of less reactive amino acids [110].

3.2.2
Dehydration in the Presence of Clays, Silica, or Alumina

Dehydration in a fluctuating environment is a well-known procedure to induce peptide bond formation (mostly with glycine) [106]. A comparison of the efficiency of different categories of clays, silica, and alumina in alternate wetting and drying cycles and a study of the nature of catalytic site allowing the formation of polypeptides has been carried out [111, 112]. Clay catalysis can also be combined with salt-induced peptide formation, leading to longer peptides [113]. The condensation also takes place with Cu(II) exchanged clays [114].

3.2.3
Reactions of *N*-Phosphoryl Amino Acids

An intramolecular phosphoric-carboxylic mixed anhydride has been proposed to explain the specific behavior of *N*-(dialkylphosphoryl)amino acids **15** that spontaneously give oligopeptides upon standing in various solvents [115]. Peptide formation was accompanied by diester exchange on the phosphoryl group via a reaction that may have been useful for nucleotide ligation.

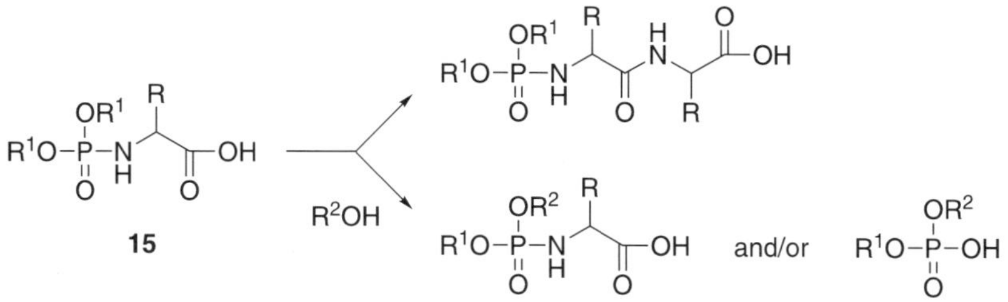

Scheme 11

The possibilities of *N*-(dialkylphosphoryl)amino acids for the prebiotic syntheses of peptides and polynucleotides have been studied in a series of papers [24, 116–122]. However, it must be emphasized that the phosphoryl group does not behave as an amino-activating group, the hydrolysis of which would be coupled to peptide bond formation. Actually, further peptide elongation requires the subsequent hydrolysis of the *N*-terminal phosphoryl group of the ligated product. In the presence of an amino acid ester, dipeptide esters **16** with an unreacted *N*-phosphoryl protection are formed, supporting the involvement of the phosphoryl group as an intramolecular catalyst rather than as an activating agent. Therefore, peptide bond formation results from a thermodynamic process taking place simply because the kinetic barrier against peptide bond formation is strongly reduced.

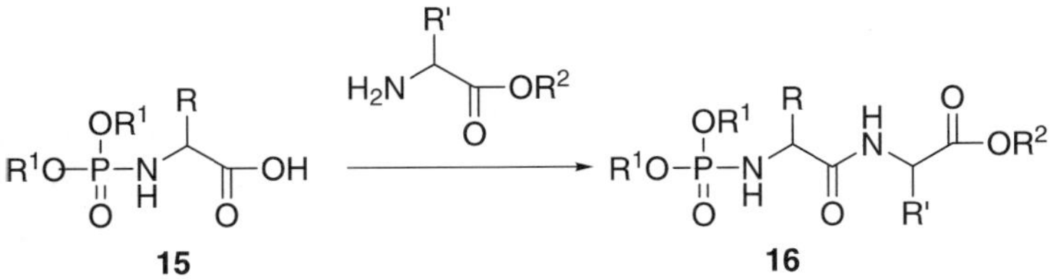

Scheme 12

Experiments with aspartic acid have shown that only α-aspartic acid dipeptides are formed [123], which is consistent with the transformation of the substrate into a pentacoordinated phosphorane mixed anhydride **17** [24] containing a five-membered ring through a self-activation process.

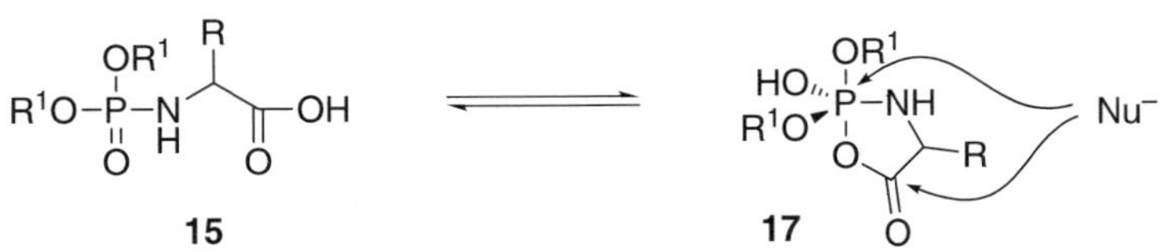

Scheme 13

This process of neighboring group participation allows the formation of a reactive intermediate by cyclization, which is in principle very similar to the concept of *overactivation by cyclization* that will be discussed more widely in the following to account for the formation of NCAs from moderately activated molecules (see Sect. 3.3). The barriers for the conversion of the cyclic phosphorane intermediate into products are lowered since the corresponding transition states are cyclic and their free energies are reduced by the preexistence of a P – N bond that is not cleaved in these processes. As a result, the conversion of *N*-phosphoryl amino acids is made easy and reversible and the formation of peptides can be driven by the reaction conditions, which must be thermodynamically favorable to bring about the free energy for peptide bond formation (low content of water and/or pH corresponding to the favorable region of Fig. 2).

Because of their abilities in both peptide and nucleic acid oligomerization, *N*-phosphoryl amino acids could have played an important role in prebiotic chemistry on condition that a plausible pathway of synthesis of these compounds is made available.

3.3 Polymerization of *N*-Carboxy Amino Acid Anhydrides (NCAs)

The interest of organic chemists in the synthesis and polymerization of NCAs has been considerable over the last century [124], with known methods for their preparation usually involving reagents such as phosgene or its surrogates and dry organic solvents. In contrast, the prebiotic relevance of NCA monomers was not accepted until 10 years ago. Nevertheless, NCAs remained very convenient monomers for polypeptide formation in studies on the origin of life, and were widely used as model compounds, e.g. in stereoselection studies [125]. Even though their capabilities as activated peptide monomers in this field had been mentioned earlier [126, 127], it is only during the last decade that several plausible pathways have been proposed for the prebiotic formation of *N*-carboxyanhydrides as well as pathways in which they are clearly identified or presumed intermediates. We will show in the following that they could even be considered as a universal activated form of amino acids. For these reasons, amino acid *N*-carboxyanhydrides deserve a particular discussion. These compounds are probably among the simplest activated derivatives of amino acids since they are in theory accessible simply by the reaction of carbon dioxide and dehydration. It is now generally admitted that the partial pressure of CO_2 in the primitive atmosphere was certainly above the present value [30, 128], and probably much higher. Therefore, the possibility of an activated peptide monomer involving a CO_2-derived moiety only is very attractive. Moreover, the presence of a mixed anhydride moiety in NCAs is not their only useful feature since they also involve a protection of the amino group, which is a striking example of atom economy since the two C and O atoms play both an activating and a protecting function. As a result, their polymerization requires a nucleophilic initiator.

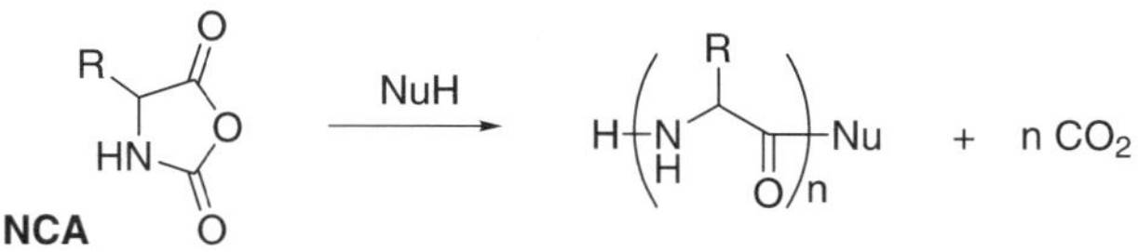

Scheme 14

The possibility of reaching high polymerization degrees (of ca. 10 in solution but which can increase to approximately 55 upon adsorption on mineral surfaces [129]) is an important difference with many other peptide monomers **18** involving a free amino group such as amino acid activated esters (X = OR′). These precursors are subject to the side-reaction yielding diketopiperazine **19** so that polymerization can be interrupted at the dipeptide stage.

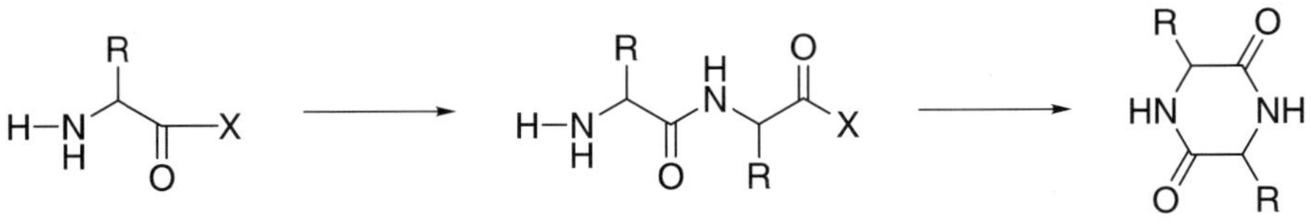

18

19

Scheme 15

3.3.1
N-Carboxyanhydrides are Stabilized Compared to other Anhydrides

The free energy of hydrolysis of NCAs has been estimated in this work to ca. $-60\,\mathrm{kJ\,mol^{-1}}$ (see Appendix). It is worth mentioning the fact that this value is less negative than those estimated in Table 1 for adenylates, *p*-nitrophenyl esters and imidazolides, which has important consequences in prebiotic chemistry. This value of $\Delta G^{\circ\prime}$ requires a careful analysis and is in relation with the unique properties of *N*-carboxyanhydrides. This lower degree of activation on a thermodynamic scale does not mean that NCAs are kinetically less reactive than adenylates or *p*-nitrophenyl esters and their kinetic reactivity is probably close to that of other anhydrides of amino acid derivatives that are thermodynamically much more unstable ($\Delta G^{\circ\prime} \sim -100\,\mathrm{kJ\,mol^{-1}}$ for amino acid symmetrical anhydride, Table 1). Actually, the relative thermodynamic stability of NCAs is related to the two different stages formally needed for building their structure: a bimolecular reaction and an intramolecular one. This peculiarity results in a higher degree of activation than could be understood from the $\Delta G^{\circ\prime}$ value, which can be analyzed by separating observed binding energies into several components [130]: the *intrinsic binding energies* ΔG^{i} (corresponding to the energy that could be obtained from the bond in the absence of entropy loss), which are additive by definition, and the *connection binding energies* ΔG^{s} derived largely from changes in translational and rotational entropy (Scheme 35). The intrinsic binding energies for the formation of carbamate ΔG^{i} (carb) and anhydride bonds ΔG^{i} (anh) are additive in NCA since both bonds are present in the NCA structure. This is not the case for *connection binding energies* ΔG^{s} because the adverse entropy to any bimolecular reaction [45, 131] has almost completely been paid for in the first step (the favorable intrinsic binding energy of carbamate formation of ΔG^{i} (carb) compensate for ΔG^{s} (carb)). Then, the second step, which is *intramolecular*, is much less entropically unfavorable (δG^{s} (anh) $\ll \Delta G^{\mathrm{s}}$ (carb)) as expected for any intramolecular reaction or cyclization involving a five-membered ring [131–133]).

In other words, the intrinsic binding energies of the two steps are additive but not the entropy losses [130], which means that the intrinsic binding free energy of the first step ΔG^{i} (carb) contributes to NCA stabilization. On the other hand, free energies corresponding to the overall entropy losses as-

Table 1 Standard $\Delta G^{o\prime}$ of hydrolysis at pH 7 and 25° for activated acetic acid or amino acid derivatives and for biologically activated phosphates reported in the literature or estimated in this work. Italicized values were estimated in this work

Compound/Component	Acetyl	Aminoacyl	Other
	$\Delta G^{o\prime}$/kJ mol^{-1} [a]		
Ester	– 19.7	*–35.1* [b]	
Phosphate	– 43.1	*∼–50* [c,d]	
Adenylate	*–54.8* [e,f]	*∼–65* [c,d,g]	
Thioester	– 31.5	*∼–47* [c]	
Imidazolide	– 54.3	*∼–70* [c]	
p-Nitrophenyl ester	– 54.4	*∼–70* [c]	
Symmetrical anhydride	– 91.2	*∼–100* [h]	
Amino acid *N*-carboxyanhydride			*∼–60* [i]
Peptide bond			– 6.0 to – 2.0[j]
ATP (to AMP & PP_i)			– 32.2
ATP (to ADP & P_i)			– 30.5
PP_i			– 33.5

[a] From [102] unless otherwise mentioned; [b] Value reported for Val-*t*RNA; [c] Estimated by adding $\Delta G^{o\prime}$(Val-*t*RNA) – $\Delta G^{o\prime}$(ethyl acetate) = – 15.4 to the value reported for the corresponding acetyl derivative; [d] Taking into account the electrostatic stabilization of the phosphate charge by the ammonium group supposed to be identical to the difference of 0.8 pK_A units between butyric acid and amino butyric acid [154] (4.6 kJ for the monanion of adenylate and 9.2 kJ for the dianion of phosphate mixed anhydride); [e] Value considered to be identical to that of acetyl phosphate monoanion at pH 7 calculated using the published value of 4.95 for the pK_A of acetyl phosphate [155]; [f] Consistent with the value of – 55.6 kJ mol^{-1} reported in [156]; [g] the differing estimation from the literature [157] of a free energy content ∼ 25 kJ mol^{-1} higher than ATP is based on the free energy of hydrolysis of *acetyl* adenylate with no correction for the presence of the protonated amino group as carried out here; [h] Estimated for an acylated AA 0.9 pK_A units more acidic than acetic acid; [i] See appendix; [j] From [101, 158]

sociated with the formation of the NCA, ΔG^s (carb) + δG^s (anh), or the linear anhydride **20**, ΔG^s (anh), must not be very different. As a result, the free energy of hydrolysis of NCA is decreased by a quantity close to ΔG^i (carb). It may represent a substantial portion of the theoretical maximal value of 38 kJ mol^{-1} [133] to 45 kJ mol^{-1} [131] as compared to acyclic anhydrides of protected amino acid **20**, which is consistent with the numerical values of $\Delta G^{o\prime}$ indicated in Table 1 for NCA ($\Delta G^{o\prime} \sim -60$ kJ mol^{-1}) and acetic anhydride ($\Delta G^{o\prime} = -91.2$ kJ mol^{-1}) or *N*-protected AA symmetrical anhydrides ($\Delta G^{o\prime} \sim -100$ kJ mol^{-1}).

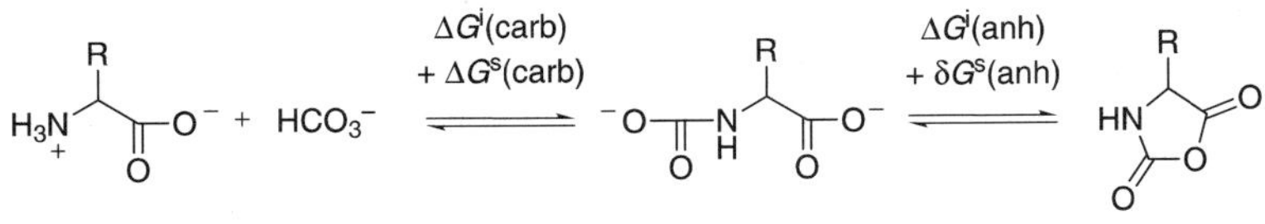

$$\Delta G^{\circ\prime}(\mathrm{NCA}) = \Delta G^{\mathrm{i}}(\mathrm{carb}) + \Delta G^{\mathrm{s}}(\mathrm{carb}) + \Delta G^{\mathrm{i}}(\mathrm{anh}) + \delta G^{\mathrm{s}}(\mathrm{anh})$$

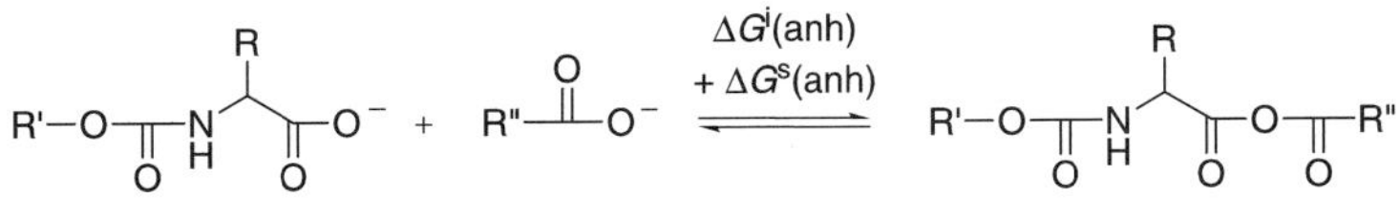

20

$$\Delta G^{\circ\prime}(\mathbf{20}) = \Delta G^{\mathrm{i}}(\mathrm{anh}) + \Delta G^{\mathrm{s}}(\mathrm{anh})$$

Scheme 16 Thermodynamic analysis of the two-stage anhydride formation leading formally to NCA compared with the intermolecular process giving the linear anhydride **20**

3.3.2 NCAs Are Kinetically as Reactive as Other Anhydrides

The acyl transfer reactivity of NCAs is not affected by this increased thermodynamic stability. This is because there is no change in the five-membered ring structure at the transition state of acyl transfer reactions, which is close to that of the tetrahedral intermediate **21**.

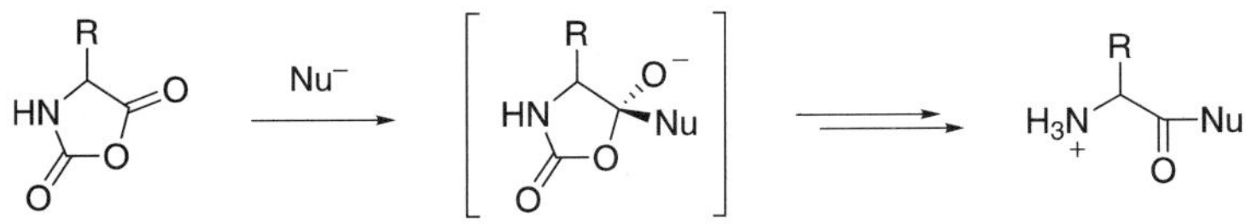

21

Scheme 17

Then, there is no reason for a difference of several orders in magnitude for this kinetic barrier compared to that of the analogous reaction of linear amino acid mixed anhydrides. Nucleophilic reactions of NCAs could even be

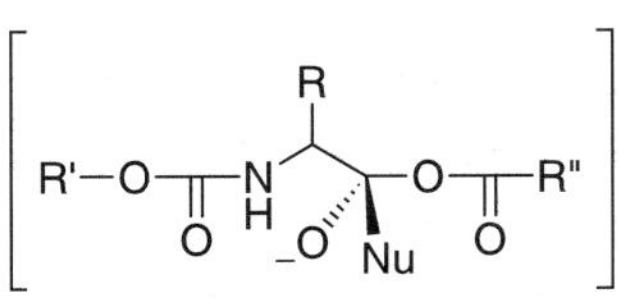

22

Scheme 18

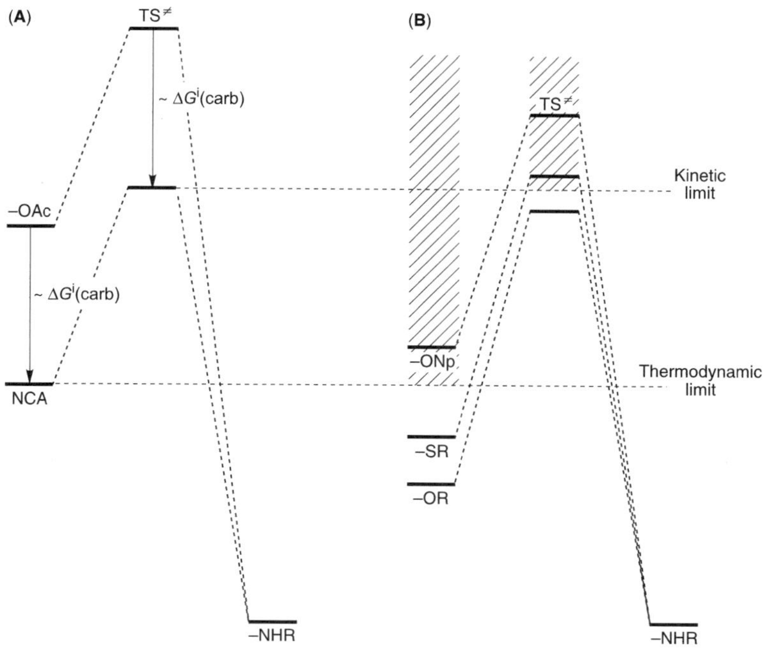

Fig. 3 Hypothetical free energy diagram proposed to account for NCA reactivity. Free energy levels of NCAs and of the transition state of their aminolysis reactions compared to those of reactions of other amino acid derivatives (AA-X). **A** Comparison with linear anhydrides (X = OAc); **B** Alkyl ester (X = OR), thioesters (X = SR) and *p*-nitrophenyl esters (X = ONp). The free energy regions corresponding to the possibility of pathways involving NCAs are shown hashed

faster since the cyclic structure is likely to minimize sterical hindrance in the tetrahedral intermediate compared to the flexible linear structure **22**.

In other words, the preserved cyclic structure of the transition state of NCA reaction ensures a degree of stabilization that is similar as the substantial one that is procured in the ground state. The free energy level of the transition state of NCA reactions can then be located below that of much less activated acyclic amino acid derivatives (Fig. 3), which results experimentally in the occurrence of CO_2-catalyzed pathways that will be analyzed subsequently.

3.3.3
NCA Formation from Less Reactive Amino Acid Derivatives

The specific stabilization of NCAs, analyzed above to be the result of their cyclic structure made of an amino acid and CO_2, suggests that their formation could be thermodynamically favorable from amino acid derivatives less reactive than anhydrides. In fact, it is known from the early work of

Wieland et al. that the hydrolysis of amino acid active esters is catalyzed by aqueous bicarbonate buffers [134] and indications of the intermediacy of *N*-carboxyanhydrides have been provided [127, 134–136]. The proposed mechanism (Scheme 36) is very similar to the catalytic pathway of α-aminonitrile hydration analyzed in Sect. 2.1.3.

On first analysis, it is tempting to consider the reaction as a drawback since the activated amino acid is converted into an NCA intermediate that is highly sensitive to the nucleophilic attack of water. Although *p*-nitrophenyl esters **18a** are clearly not likely primitive activated peptide monomers, their reactivity can be considered as a model of that of compounds bearing a similar degree of activation. A closely related observation has indeed been reported for aminoacyl adenylates **18c** that are quickly hydrolyzed at concentrations of bicarbonate buffers as low as 1 mM [137, 138], which is clearly meaningful since these mixed anhydrides are involved as intermediates in protein biosynthesis. It is also clear that the transient formation of NCAs in these processes can have several advantages, namely, the suppression of diketopiperazine formation, which may lead to longer peptide chains, and a higher sensitivity to nucleophiles so that hydrolysis would not be favored compared to nucleophilic attack and rates may be increased. An observation already noticed by Brack [127] supports the last conclusion: reports of peptide syntheses conducing to high polymerization degrees that were carried out in media reproducing prebiotic environments involved aqueous bicarbonate buffers [139, 140] so that the intermediacy of NCAs is highly probable.

There is a striking similarity between the CO_2-promoted catalytic pathway (Scheme 36) and the usual representation of enzymatic catalysis. It is because both processes are partly based on the concept of *induced intramolecularity* [44]. Except the nature of binding, there is indeed no fundamental difference between the utilization of the intrinsic binding energy from the

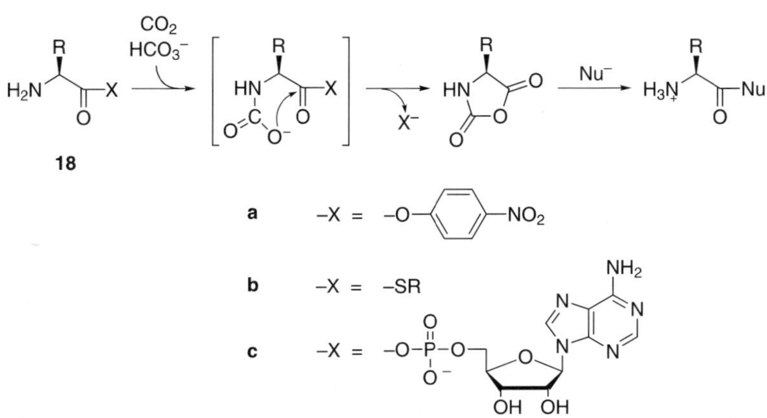

Scheme 19 Carbon dioxide-catalyzed hydrolysis of activated amino acid derivatives

covalent bond with CO_2, ΔG^{i} (carb), to lower the free energy level of the transition state (Fig. 3) and the utilization of *non-covalent interactions* with non-reactive portions of substrates that has been proposed [45] to be responsible for an essential part of enzyme power. In this way, carbon dioxide can be considered as belonging to the class of small-molecule catalysts that played a role of protoenzymes [44] at the beginning of chemical evolution.

3.3.4
NCA Formation from Reactive Amino Group Derivatives

NCAs are not only produced from acyl group activation in amino acids, they are also obtained from amino acid derivatives **23** with amino groups under the form of carbamates with good leaving groups (Scheme 37). Since carbamates with good leaving groups or related carbamic acid derivatives are easily converted into isocyanates [141, 142] there is a possibility of transfer of activation to the acyl group by cyclization of the isocyanate intermediate **24** into NCA (see below the paragraphs on CDI and COS activation, for instance). But there is also a possibility of direct nucleophilic substitution avoiding the isocyanate intermediate (Scheme 37, path b).

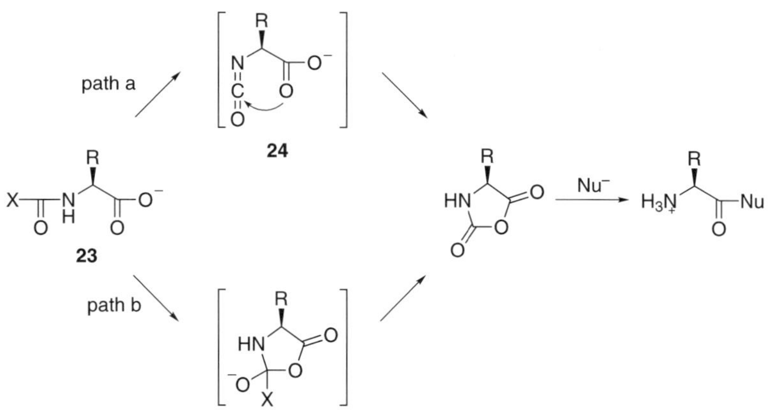

Scheme 20 Intramolecular transfer of activation from amino group derivatives

3.3.5
NCA as the Most Activated Form of Amino Acids in an Aqueous Prebiotic Environment

The fast and easy conversion of activated amino acids **18** with a $\Delta G^{\circ\prime}$ above the thermodynamic limit of Fig. 3 into NCA and the similar conversion of *N*-substituted derivatives **23** involving good leaving groups show that most extremely activated amino acid derivatives must have been converted into NCAs in bicarbonate-buffered aqueous solution at moderate pH (5.5–8.5).

This unexpected consequence of the efficient reaction of carbon dioxide can be expressed in an other way: NCAs can be considered as the most activated amino acid species achievable in water in the environment of the primitive Earth. The only exception would be species bearing a chemical protection of the α-amino groups that are unlikely because peptide elongation would have been complicated by the necessity of an additional deprotection step. From a prebiotic perspective, there is consequently no need to search for activated amino acid derivatives with a degree of activation higher than NCAs (thermodynamic limit in Fig. 3).

3.3.6 NCAs as Intermediates in the Reactions of Much Less Activated Amino Acid Derivatives

But NCAs are not only involved as thermodynamically favored intermediates because of their relative stabilization. They can also be involved as kinetically competent intermediates in the reaction of species with an energy content lower than NCAs (thermodynamically more stable) so that only a small degree of conversion would be expected to take place at equilibrium (Fig. 3). This seems to be the case for thioesters ($\Delta G^{\circ\prime} \sim -47\,\mathrm{kJ\,mol^{-1}}$), for which the presence of bicarbonate buffers has been shown to strongly increase the polymerization degree [127]. One of the explanations of the difference with methyl esters, which do not undergo a carbon dioxide-catalyzed reaction [127], is that the free energies of the transition states for methyl ester aminolysis or hydrolysis could lie below that of the NCA reaction (kinetic limit in Fig. 3), therefore no catalysis by CO_2 could take place. The CO_2-catalyzed pathway is a consequence of NCA reactivity, which is equivalent to that of species with a $\Delta G^{\circ\prime}$ value 30–40 $\mathrm{kJ\,mol^{-1}}$ more negative than that predicted by the actual thermodynamic value ($\Delta G^{\circ\prime} \sim -60\,\mathrm{kJ\,mol^{-1}}$). Taking advantage of this principle of *overactivation by cyclization*, NCAs can be involved as intermediates in the reaction of much less activated species *provided that the level of the transition state of the uncatalyzed reaction is above that of the CO_2-catalyzed reaction.* It must be emphasized that there is no lower limit for the free energy content of the amino acid derivative in the ground state so that the CO_2-catalyzed pathway might in principle be observed for unactivated species provided that the requirements for the transition state are fulfilled.

The first important consequence of these thermodynamic and kinetic analyses for the early synthesis of peptides is that NCAs are likely intermediates of the process whatever could have been the actual chemical precursor of these unusually activated derivatives. The second one is that derivatives with an unexpectedly low reactivity may have been involved as NCA precursors provided that two conditions are fulfilled, namely, first their uncatalyzed hydrolysis must be slow and second they can be converted into NCAs in the presence of CO_2. As a result, studies of uncoded peptide bond formation are founded in using NCAs as substrates (even when obtained as the reactive species of the non-prebiotic CDI activation) for mimicking prebiotic processes. Further studies on the emergence of the genetic code would then preferentially select activation under the form of NCAs although other pathways could never be excluded.

3.3.7 New Pathways for NCA Prebiotic Synthesis

Nitrosation of CAAs

Commeyras and co-workers proposed for the first time a prebiotically relevant synthetic pathway [143, 144]. They established that the nitrosation of *N*-carbamoylamino acids **25** (otherwise believed as unreactive in prebiotic environments) by nitrogen oxides (typically obtained by mixing O_2 and NO) quantitatively releases NCA. Although first tested in organic solvent [143], the reaction proved to work in the solid state [144], and also in aqueous solution [145], where the NCA can be observed for ca. 1 h prior to its conversion into either free amino acids or peptides.

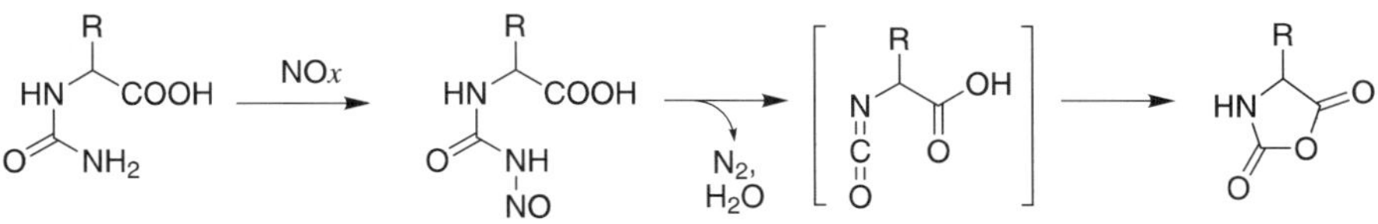

25 **23a** **24**

Scheme 21

The reaction involves the formation of an N'-nitrosourea **23a**, then its decay into nitrogen, water and a transient isocyanate **24** that readily cyclizes onto the vicinal free carboxy group. Besides the involvement of an activated nitrogen oxide species, the main driving force in this pathway is the fragmentation of the N'-nitrosourea into inactivated molecules (N_2, H_2O). An alternate (and probably slower) mechanism is the intramolecular nucleophilic attack of the N'-nitrosourea **23a** by the vicinal carboxy group. *N*-substituted NCAs may also be formed by this latter pathway, even though the corresponding α-carboxy isocyanate is not accessible. This has been confirmed by the formation of proline-NCA from *N*-carbamoyl proline [145].

The prebiotic relevance of this pathway was examined in several papers [144, 146, 147]. The main requirements of such a pathway are (*i*) the presence of CAA in the prebiotic pool, (*ii*) the presence of sufficient amounts of NO_x in the atmosphere and, (*iii*) the possibility of a concentration process, e.g. through water evaporation.

(i) The prebiotic formation and presence of CAA has been discussed previously [43] (see Sect. 2.1.4).
(ii) The reaction requires NO_x (NO_2, N_2O_3, N_2O_4) species, which spontaneously form from mixtures of either NO and $O^•$, NO and O_2, or NO and NO_2. Although the primitive Earth atmosphere was very probably mostly made of N_2 and CO_2, NO_x species may have formed as minor compounds after either thermal or photochemical activation of N_2-containing gas mixtures. Thermal activation (mostly through lightning) is due to either volcanic activity [148, 149] or meteoritic impacts. Modeling of primitive Earth atmosphere photochemistry showed that these NO_x species may have been present in several configurations of $CO_2/N_2/H_2O$ mixtures, however in moderate steady-state concentration. Under such conditions, impacts of medium-sized bodies can however have transiently raised the NO abundance by several orders of magnitudes [147].
(iii) The concentration process (e.g. by evaporation from a lagoon or a tidal shore) is necessary to ensure an efficient exposure of CAA to NO_x from the atmosphere, while avoiding premature hydrolysis of NCA. Indeed the nitrosation reaction generates acidic conditions that are unable to ensure NCA conversion into peptides. The contact of newly formed NCA in an aqueous medium of pH > 4 (e.g. sea water) is then necessary to observe the subsequent formation of peptides [147]. This requires the presence of both liquid water and emerged land at the surface of the primitive Earth.

Commeyras and co-workers investigated the kinetics of NCA polycondensation in water [150]. Since the aminolysis of the NCA is much faster than its hydrolysis, the formation of peptides is possible provided that sufficient concentration of free amino groups is attained. Practically the formation of peptides is observed at any pH above 4. In the pH 4–7 range, the competition with hydrolysis enhances the effects of chemo- and stereo-selectivity of NCA condensation, through removing unreacted NCA (for example, the less reactive enantiomer in excess) before complete conversion.

Reactions of Diacyldisulfides

While thiocarboxylic acids themselves are not effective acylating reagents in aqueous solution, the diacyldisulfides **26**, that are formed when thioacids are oxidized, are extremely effective [151].

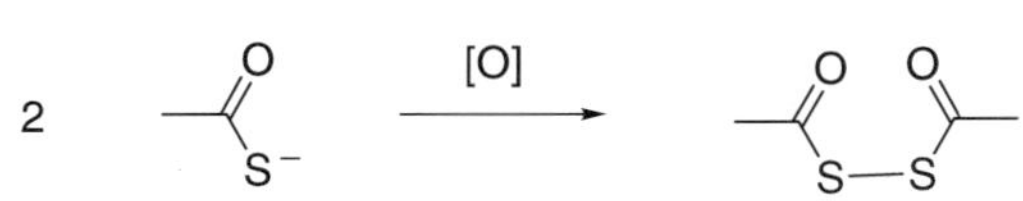

26

Scheme 22

The process is also effective with amino acid derivatives **27** as shown by the reaction of thioglutamic acid with an oxidizing agent, but it is very slow [152]. The addition of bicarbonate to the reaction mixture greatly accelerates the reaction suggesting that the *N*-carboxyanhydride of glutamic acid is an intermediate in the accelerated reaction.

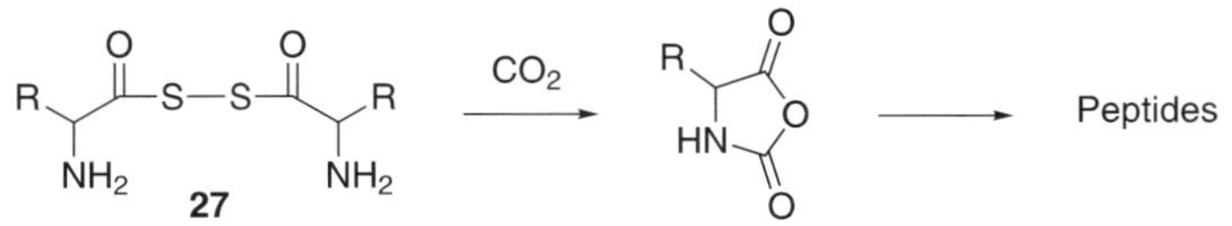

Scheme 23

Activation by Carbonyl Sulfide

COS that is presently found in volcanic gases has been proposed as a prebiotic activating agent for amino acids [153]. Amino acids in solution would have readily reacted with COS to give thiocarbamates **28** that can be converted into NCAs in significant yields.

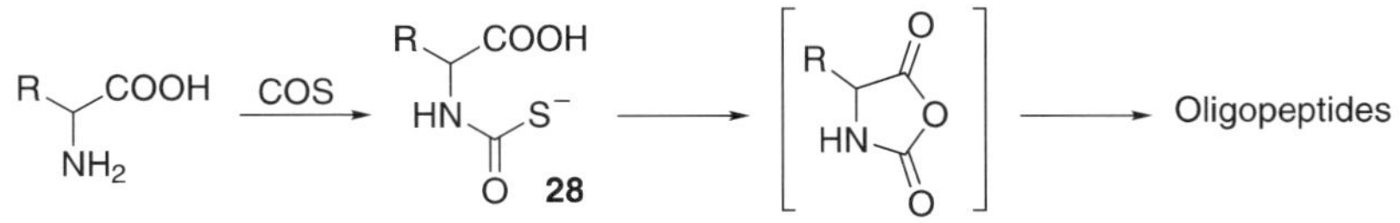

Scheme 24

These yields can be improved if an oxidizing agent is present, which allows the reaction to proceed through disulfide **29** in a way that is similar to the process indicated above for diacyldisulfides.

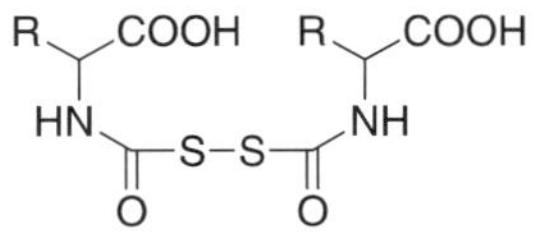

29

Scheme 25

Metallic ions were shown to be capable of replacing oxygen in the oxidation step.

3.4 Polymerization of other Activated Amino Acids

The standard free energy of hydrolysis of biomolecules at pH 7 and 25 °C is generally used in biochemistry textbooks as a parameter to characterize their level of activation [102]. However, only very few values are available for activated amino acid derivatives and any comparison of peptide monomers is impossible in this context. We collected available values for activated amino acids and attempted an assessment of free energies of hydrolysis for *p*-nitrophenyl esters, imidazolide, and the mixed anhydrides with phosphate and adenylate starting from the values reported for acetyl derivatives. These values are displayed in Table 1. Although the values correspond to 1 M standard state of reactants and products and some of them have some degree of uncertainty, the tendency is clear enough to get useful conclusions in the field of prebiotic chemistry.

3.4.1 Amino Acid-Phosphoric Acid Mixed Anhydrides

Among activated forms of amino acids, mixed anhydrides with inorganic phosphate or phosphate esters require a special discussion because they are universally involved in peptide biosynthesis through the ribosomal and non-ribosomal pathways. These mixed anhydrides have stimulated studies in prebiotic chemistry very early in the history of this field. Amino acyl adenylates **18c** have been shown to polymerize in solution [159, 160] and in the presence of clays [139]. However, their participation as major activated amino acid species to the prebiotic formation of peptides from amino acids is unlikely for at least two reasons. Firstly, amino acid adenylates that have a significant lifetime in aqueous solution become very unstable as soon as either CO_2 or bicarbonate is present at millimolar concentration [137]. Lacey and co-workers [161] were therefore conduced to consider that CO_2 was absent in the primitive atmosphere for aminoacyl adenylate to have a sufficient lifetime and then to allow for the emergence of the modern process of amino acid activation and of the translation apparatus. But this proposition is unlikely, as shown by the analysis of geological records in favor of CO_2 contents in the atmosphere higher than present levels [128]. It is also in contradiction with most studies of the evolution of the atmosphere of telluric planets [30, 32]. Secondly, there is no prebiotic pathway available for adenylate formation and ATP proved to be inefficient in this reaction [162].

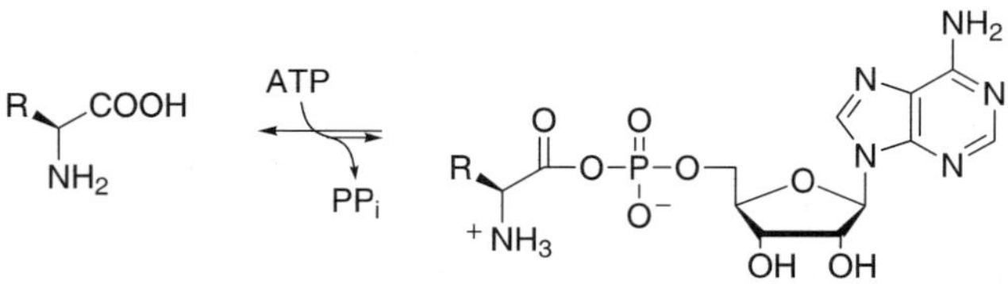

18c

$\Delta G \sim 33$ kJ mol^{-1}

Scheme 26

Our analysis of the free energy of hydrolysis of potentially prebiotic amino acid derivatives shows that the energy content of the phosphoanhydride bond of ATP is lower than that of the adenylate anhydride **18c** by *ca.* 33 kJ mol^{-1}. The reaction is highly unfavorable and, at equilibrium, less than 1 molecule over 10^5 would be under the form of the aminoacyl-adenylate mixed anhydride at pH 7. Actually, in biosynthetic pathways [157, 163], adenylates mostly remain in a bound state ($K_D \sim 10$ nM for isoleucyl-*t*RNA synthetase) and are stabilized in the active site of the adenylate-forming enzyme so that bound reagents and products are close to equilibrium. Moreover, to drive adenylate formation to completion, the process can be coupled to pyrophosphate hydrolysis as a result of inorganic pyrophosphatase activity [164], which allows the system to take advantage of the free energy of both phosphoanhydride bonds of ATP. But a prebiotic process of that kind would have required two different but efficient prebiotic catalysts, the first one to achieve adenylation, and the second one for hydrolyzing pyrophosphate whereas being inactive with respect to ATP or other polyphosphates. Therefore, adenylation with ATP or even the formation of other mixed anhydrides from phosphoanhydrides can be ruled out as abiotic processes for amino acid activation.

Alternatively, the difference in free energy between NCAs and adenylates, which is approximately of 5 kJ mol^{-1} would have allowed the presence of small but substantial amounts of phosphate mixed anhydrides (possibly adenylates) of amino acids at equilibrium.

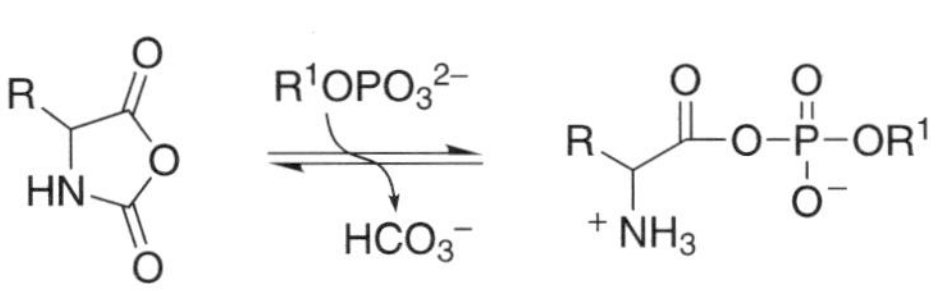

18d: R^1 = H or ⊖ ($\Delta G \sim -10$ kJ mol^{-1})

18e: $R^1 \neq H$ ($\Delta G \sim 5$ kJ mol^{-1})

Scheme 27

This possibility has recently been supported by the observation of a reaction of inorganic phosphate (R^1 = H) with valine NCA yielding the mixed anhydride **18d** (R^1 = negative charge) as a reaction intermediate [165]. The reaction of inorganic phosphate is apparently complete in agreement with the fact that aminoacyl phosphates are predicted to display a free energy level 15 kJ mol^{-1} lower than that of adenylates (Table 1). Surprisingly, the mixed anhydride turned out to behave as an activated phosphate instead of an activated amino acid giving rise to a process in which the hydrolysis of NCA is coupled to the free energy transfer to phosphate, so that NCAs may have been involved not only in prebiotic peptide synthesis but also in phosphorylation processes.

Starting from phosphate monoesters ($R^1 \neq H$), including nucleotides, the mixed anhydrides would be less easily observable from NCAs since the equilibrium is slightly unfavorable as predicted from the values of Table 1. But, in this case, the phosphoryl transfer pathway is unlikely to be predominant with the monoester mixed anhydrides **18e**. The latter have indeed been shown to react as aminoacylating agents [166, 167], which is consistent with the biological role of adenylates. Although it is a likely possibility that is currently investigated in our group, the formation of aminoacyl phosphate mixed anhydrides from NCAs and unactivated nucleotides under prebiotically pertinent conditions and its consequences for primitive peptide formation remains to be supported by published experiments.

In relation with the unique reactivity of *N*-(dialkylphosphoryl)amino acids mentioned above (see Sect. 3.2.3), a conversion of the carboxylic-phosphoric mixed anhydride **18f** ($R^1 = R^2 =$ alkyl) would provide an appealing pathway for the formation of the *N*-phosphorylated derivatives. The reaction of NCAs might have then allowed a pathway for the formation of these versatile intermediates that can be useful for nucleotide ligation.

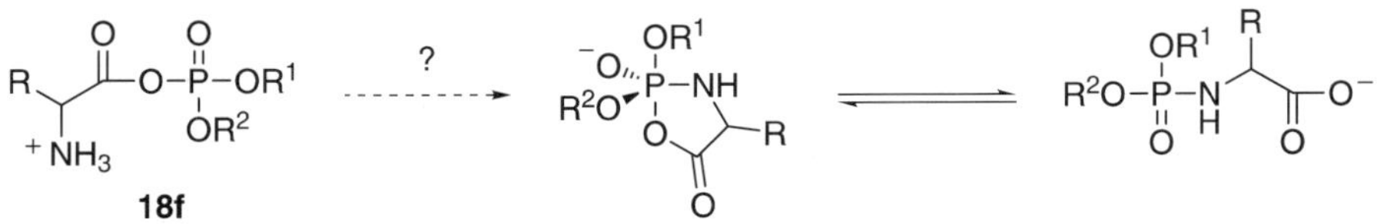

Scheme 28

However, the possibility that aminoacyl phosphoanhydrides involving phosphate diesters may have been formed from NCAs is unlikely. The absence of a negative charge on the phosphate moiety would render acyl and phosphoryl groups more sensitive to nucleophilic attack because of the loss of electrostatic repulsion with the nucleophile. As a result, the free energy content of these compounds at pH 7 would reach a level approximately 35 kJ mol^{-1} above that of the corresponding monoester derivatives **18e** assuming a free energy level similar to that of the undissociated conjugated acid **18f** ($R^1 =$ alkyl, $R^2 =$ H) and a pK_A value of 1 for the acid dissociation constant of its acylated phosphoric acid moiety. The formation of anhydride **18f** in significant concentration is thus unlikely.

Several systems based on the potentialities of amino acid–phosphoric acid mixed anhydrides have been devised to check the idea that the genetic code developed from an early pathway of RNA-dependent peptide synthesis in an RNA world [168]. RNA sequences have thus been selected that are capable of self-aminoacylation using amino acid adenylates, catalyzing a reaction chemically similar to the aminoacylation of *t*RNA by the protein aminoacyl *t*RNA synthetases [169].

Thanks to the fact that the reaction of a nucleoside 5′-triphosphate with amino acids is less unfavorable at low pH, 5′-triphosphate RNAs capable of self-aminoacylation have been selected from a randomized pool of sequences (120-mers) at pH 4–4.5 [170]. The selection process involved the selection and amplification of the thiol-containing sequences formed by reaction of mercaptopropionic acid with the 5′-terminal triphosphate group.

Scheme 29

In addition to the reaction of mercaptopropionic acid, mixed anhydrides were also formed and identified starting from leucine and phenylalanine in the presence of Ca^{2+} ions, showing that RNAs can replace protein aminoacyl *t*RNA synthetase catalysts for amino acid activation. The formation of a detectable amount of aminoacyl 5′-phosphate polynucleotide seems to be in contradiction with the instability predicted for aminoacyl adenylates (Table 1), however it can be explained by the low pH value increasing their stability and the fact that the selected RNA structures are likely to stabilize the mixed anhydride moiety of the covalent conjugate by favorable intramolecular interactions induced by folding.

Recently, the formation of a peptide bond has been observed in a system inspired by the contemporary process that uses aminoacyl (mononucleotide) adenylates as intermediates for aminoacyl *t*RNA synthesis [171]. Thus, a system made of three components (Fig. 4), an aminoacyl (ester) mini-

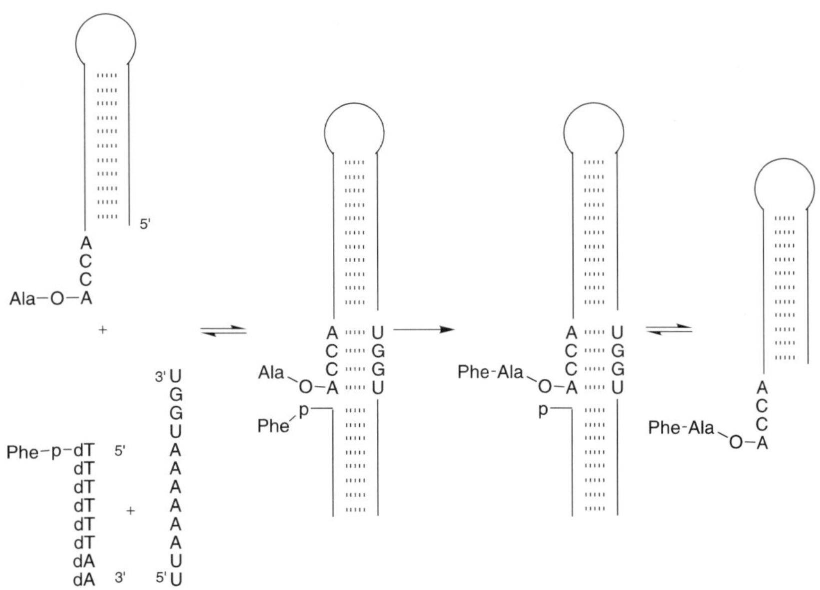

Fig. 4 Peptide bond formation by reaction of an aminoacyl (anhydride) oligonucleotide as described in [171]

helix, a highly reactive aminoacyl (anhydride) phosphate oligonucleotide, and an RNA guide was shown to be efficient at much lower concentrations than an intermolecular system [171].

A similar system was also capable to promote acyl transfer allowing the synthesis of a 2′,3′-aminoacyl (ester) oligonucleotide from an aminoacyl (anhydride) 5′-phosphate oligonucleotide [172].

3.4.2 Amino Acid Thioesters and Amino Thioacids

It has been devised that the thioester bond served as the first high-energy bond in a "thioester world" proposed to have occurred as an intermediate stage between an abiotic world and the RNA world [173]. This hypothesis is supported by the existence of many pathways involving thioesters as intermediates in the metabolism of modern living organisms such as non-ribosomal peptide synthesis and coenzyme A-thioester-linked reactions [174, 175]. Another indication supporting a prebiotic role for thioacids and thioesters is the fact that sulfur compounds at low oxidation states were likely to be present due to an increased volcanic and hydrothermal activity. As, under reducing conditions, aminonitriles are likely to have been formed in substantial amounts [1, 43], their reaction with H_2S would have produced thioamides and thioacids [151]. It is also supported by the free energy content of thioesters (Table 1) that is consistent with a possibility of conversion of inorganic phosphate into aminoacyl phosphates and then with a prebiotic phosphorylation process [173]. On the other hand, the formation of aminoacyl adenylates (or other mixed anhydrides with phosphoric acid monoesters) from aminoacyl thioester is not favorable (Table 1) unless a prebiotic catalyst playing a role similar to that of modern adenylate-forming enzyme would be present to stabilize the mixed anhydride.

The possibilities of amino acid thioesters in peptide synthesis have been demonstrated very early as well as the effect of carbon dioxide supporting the involvement of *N*-carboxyanhydrides in the hydrolysis and polymerization pathways [127, 176, 177]. Aminothioacids can also be converted into NCAs [152, 178, 179] but they can, in addition, be activated by oxidation into disulfides which are much more reactive [151].

Weber [61, 62] has developed in the context of prebiotic chemistry an original pathway for α-aminothioester synthesis [180], which can start from hydroxyaldehydes **30** intermediates in the formose reaction (a likely prebiotic pathway to carbohydrates). Obviously, thioesters themselves are not observed as products because of their fast hydrolysis in the medium, but they could be converted into peptide bonds in the presence of amino acids or peptide free amino groups, and into mixed anhydride with phosphoric acid in the presence of inorganic phosphate. The reaction involves two key-steps: the condensation of ammonia and of the mercaptan on α-keto aldehyde **31** (e.g. resulting of the dehydration of an aldol-condensation product), then the redox rearrangement of the resulting imino hemithioacetal **32**. Its main originality lies in the fact that the free energy present in formaldehyde (the monomer precursor of formose products) is converted into valuable carbohydrate or peptide products.

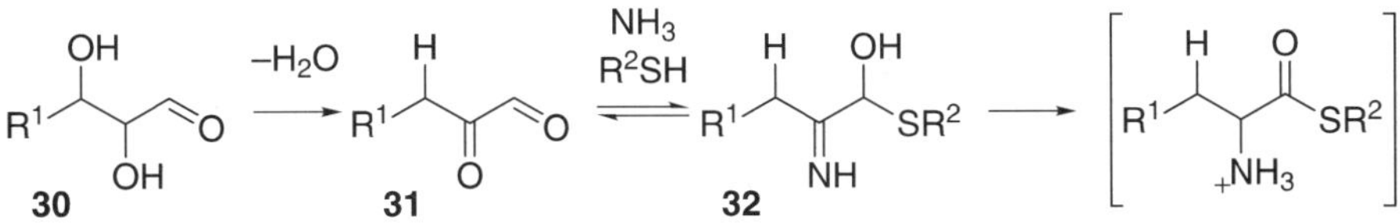

Scheme 30

3.4.3 Other Amino Acid Derivatives

As reminded by Lazcano and Miller [181], a possibility for the formation of prebiotic polymers is the spontaneous synthesis of a polymer from high-energy precursors. The direct polymerization of glycine nitrile into polyglycine [182–184] is sluggish, but it is thermodynamically favorable and the possibility of indirect catalytic processes remains to be explored.

3.5 Activating Agents

3.5.1 *N*, *N′*-Carbonyldiimidazole (CDI)

Since *N*,*N′*-carbonyldimidazole has been shown to induce peptide bond formation in aqueous solutions of amino acids [185], this way of activation has been repeatedly used in the field of prebiotic chemistry though CDI could hardly be considered as a prebiotic activating agent. During the last years this method of activation has been used to study the stereoselectivity of the reaction in solution [186–188], in aggregates [189] and the polymerization on solid minerals [129, 190]. The initial intermediate that accumulates after CDI activation of an amino acid in a concentrated imidazole buffer (pH 7) is an *N*-(imidazol-1-yl-carbonyl)amino acid, **23b**, which is then converted into oligopeptides.

Intermediate **23b** can be obtained by a direct nucleophilic attack of the uncharged amino group (present at low equilibrium concentration at pH 7) but the mechanism of this step may be more complex as shown by the observation of a catalytic effect of CO_2 in amidations using CDI [191]. Moreover, the reaction of the carboxylate group followed by an intramolecular acyl transfer is also a possibility. CDI-promoted peptide formation was shown to display

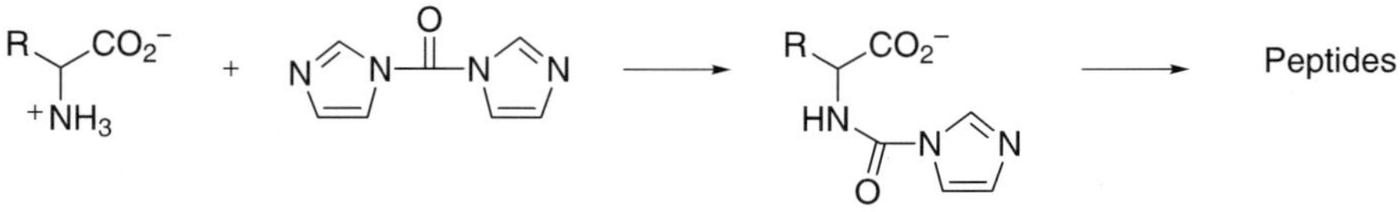

23b

Scheme 31

unusual features, namely the absence of diketopiperazine side-product and the fact that dipeptides are not polymerized by CDI activation in water [185]. These features are consistent with a polymerization mechanism involving low concentrations of an *N*-carboxyanhydride intermediate as mentioned above for the conversion of carbamate derivatives of aminoacids with good leaving groups. The conversion of carbamylimidazole intermediate **23b** into NCA is possible because of the easy elimination of imidazole giving isocyanate **24**, but the fact that a similar process was also observed with the *N*-methylated compound derived from sarcosine shows that acyl transfer via an addition-elimination mechanism (path b of Scheme 37) is also possible [126].

3.5.2 Other Activating Agents

Water-soluble carbodiimides have been used to polymerize negatively charged β- and α-amino acids [192, 193]. But it turned out that, starting from α-amino acids, high polymerization degrees can only be obtained with aspartic acid using these activating agents. Possibly the participation of the β-carboxyl group allows the transient formation of anhydride **33**. However, carbodiimides are usually not considered as prebiotic reagents though it must be noticed that unsubstituted carbodiimide can be formed by tautomerization of cyanamide in the presence of water at low temperature, which may have important implications for both cometary and interstellar chemistry [194].

Carbon monoxide has also been shown to act as an amino acid condensing agent at 100 °C in the presence of precipitated (Ni,Fe)S and of thiols or hydrogen sulfide [195].

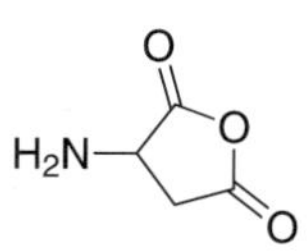

33

Scheme 32

4 Protometabolisms

Several processes involving the continuous feeding in activated reactants supplying chemical energy have been proposed to lead to the emergence of proto-metabolisms (otherwise called chemo-metabolisms), defined as a sequence of thermodynamically favorable chemical reactions (usually cyclic) through which more evolved species could have been produced, and that could have been the starting point from which life developed.

4.1 Systems Based on Amino Acid and Peptides Only

The Primary Pump Scenario

From the constraints of *N*-carbamoyl amino acid activation with NO_x (see above Sect. 3.3.7), Commeyras et al. designed a scenario they called the *Primary Pump*, which might have taken place on tidal shores during the Hadean [146]. This scenario involves several chemical and physical steps (Fig. 5), and relies on the constant income of AA and/or CAA, and on the recurrent alternation of dry and aqueous periods. These are necessary to ensure NO_x-mediated CAA activation into NCA, then to provide adequate conditions for NCA condensation. The driving force is provided by the continuous income in activated species (NO_x, cyanate) as well as by phase changes (drying, then dissolution after watering). Slow competing reactions, namely peptide hydrolysis and epimerization allow the peptide pool to evolve through chemical selection. This allows in theory the selection of privileged sequences, e.g. of regular tacticity. If the time elapsed between each drying/wetting cycle is short enough so that peptide hydrolysis is not complete, peptides can in principle be accumulated in the medium.

This *Primary Pump* scenario is a model of protometabolism [147]. Its features served to build a theoretical model (limited to dipeptides, Scheme 38) by which homochirality could have emerged from a racemic amino acid world without needing autocatalysis [196]. Actually, this model shows that stereoselectivity at three different stages (with corresponding selectivity ratios in brackets): NCA coupling with AA (α), dipeptide hydrolysis (β), and dipeptide epimerization (γ) is enough to promote homochirality. The racemic composition is not stable for certain values of the set of selectivity ratios α, β, γ (taken as parameters) and provided the system is supplied with chemical energy to continuously recycle the amino acid into NCA.

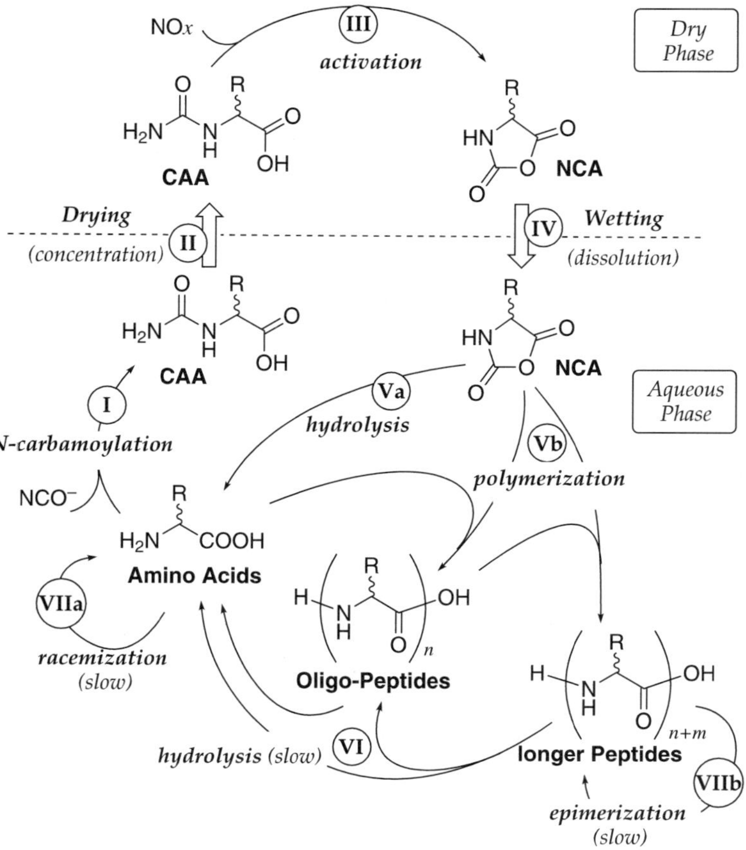

Fig. 5 The *Primary Pump*, a peptide-based protometabolism scenario [146] involving the following steps: **I:** amino acid *N*-carbamoylation; **II:** concentration through drying; **III:** NO_x-mediated CAA activation; **IV:** dissolution through watering (by e.g. sea water); **V:** NCA reaction in aqueous phase (**Va:** NCA hydrolysis; **Vb:** condensation with AA or peptide); **VI:** slow hydrolysis of peptide bonds; **VII:** α-carbon epimerization (**VIIa:** of amino acid and CAA; **VIIb:** of peptide residues). Additional steps corresponding to peptide *N*-carbamoylation/nitrosation have not been mentioned for the sake of clarity. It is worth mentioning that although the *N*-carbamoylation of peptides renders them unreactive towards NCA, this is reverted by NO_x-mediated nitrosation [197], thus keeping peptides within the polymerization process

Carbon Monoxide-driven Peptide-cycle

Extending a previous report of amino acid activation by CO in the presence of colloidal (Ni,Fe)S [195], Wächterhäuser and co-workers have recently proposed a primordial peptide cycle in which CO drives peptide synthesis and cleavage in the presence of colloidal (Ni,Fe)S [198]. The reactions were car-

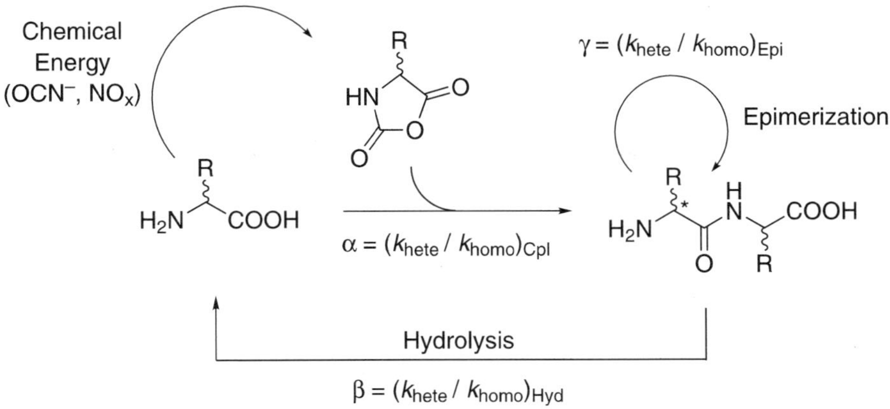

Scheme 33 Theoretical model for the emergence of homochirality avoiding autocatalysis as proposed in [196]. k_{hete} and k_{homo} represent the rate constants of "homochiral" and "heterochiral" reactions respectively

ried out by heating solutions of amino acids at 100 °C under an atmosphere of CO (1.05 bar). Under these conditions α-amino acid *N*-carboxyanhydrides have been proposed to be formed through a reaction of CO with free amino acids at the surface of the catalyst and then to be responsible for peptide elongation at the *N*-terminus of a peptide chain ($H_2N-Pept^C$). Peptide degradation is proposed to involve the reaction of CO leading to the cleavage of the *N*-terminal residue through hydantoin and urea intermediates (Scheme 30).

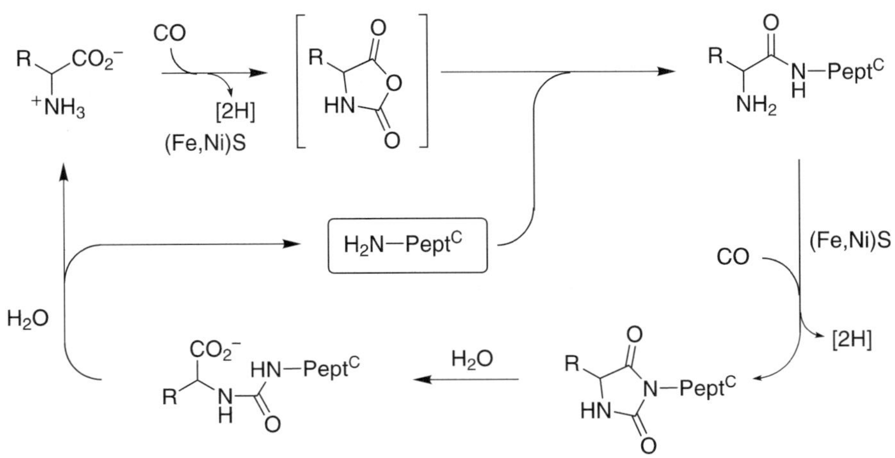

Scheme 34

4.2
Interactions between Amino Acids, Peptides, and Sugars

The process called "the sugar model" [61] by which formose reaction intermediates (glycoaldehyde, glyceraldehyde) can be converted into valuable aminothioester intermediates has been mentioned above. However, Weber has extended the connection between formose reaction and amino acid chemistry by the detection of a catalytic effect of amino acids [62]. The system can then yield a large variety of biomolecules including sugars, acetaldehyde, glyoxal, pyruvate, imidazole, hydroxy- and amino-thioester and then peptides. The experimental pH value of 5.5 is not far from the value expected under an atmosphere with a high CO_2 content, which supports the relevance of the reaction to the conditions of the primitive Earth. Amino acid catalysis proved to be stereoselective showing that L-amino acids preferentially yielded D-threose [199]. Importantly, the effect was especially noticeable for α-methyl amino acids that have been observed in non-racemic ratio in Murchison meteorite [69].

4.3
Did Biochemical Amino Acid Activation Evolve from an Early Nucleotide Activation Pathway?

We can also get some information by comparing the modern biosynthetic pathways to the capabilities of prebiotic chemistry. Amino acids are usually activated in living organisms by reaction with ATP both through the ribosomal and non-ribosomal peptide synthesis pathways. Amino acyl *t*RNA synthetases bind ATP and free amino acids to cause the highly unfavorable adenylate anhydride formation to be close to equilibrium in the active site.

$$\mathrm{AA} + \mathrm{ATP} \underset{}{\overset{K}{\rightleftharpoons}} \mathrm{AA{-}AMP} + \mathrm{PP_i}$$

$K \sim 1$ for enzyme-bound substrates and products

$K \sim 10^{-6}$–10^{-5} at pH 7 in water

Scheme 35

Regarding the origin of life problem, it remains to determine why this process has been selected by evolution whereas nearly no mixed anhydride is present at equilibrium in the absence of enzyme. If we assume the hypothesis that life emerged from a coevolution process involving both amino acid and nucleotide chemistries, it follows that the translation apparatus must have been formed very early. The only possibility that remains to explain the selection of highly unstable intermediates in peptide biosynthesis is to consider that aminoacyl adenylates were first formed from an alternative mechanism. Then, they were conserved during further evolution because their free energy level had been reduced by favorable interactions with aminoacyl adenylate binding proteins (or RNAs) that became the ancestors of modern adenylate-forming enzymes but played initially a different role. In relation with the data supporting a prebiotic role for NCAs presented in this review, the hypothesis can be made that aminoacyl adenylates could have been formed by reaction with *N*-carboxyanhydrides. According to our estimate, there is a 5 kJ mol^{-1} difference only in standard free energy between NCAs and adenylates so that a minor but significant concentration of adenylate may be formed through a process that takes place spontaneously in the absence of catalyst as shown by the similar reactivity observed for the reaction of NCAs with inorganic phosphate [165].

This hypothesis on the formation of aminoacyl adenylates or other amino acid-nucleotide mixed anhydrides let us put forward a scenario (Scheme 39) consistent with the ideas that life arose through a process involving a metabolic linkage between peptides and nucleic acids. This scenario is quite naive because it is based on the biochemistry of present day living organisms and evolution may have followed more complex pathways. It requires a complete reversion of the rule learned from biochemistry that activated amino acids are only building blocks in protein synthesis while nucleotides play many other roles in the cell than simply being nucleic acid monomers [200] and very importantly the role of energy carriers under the

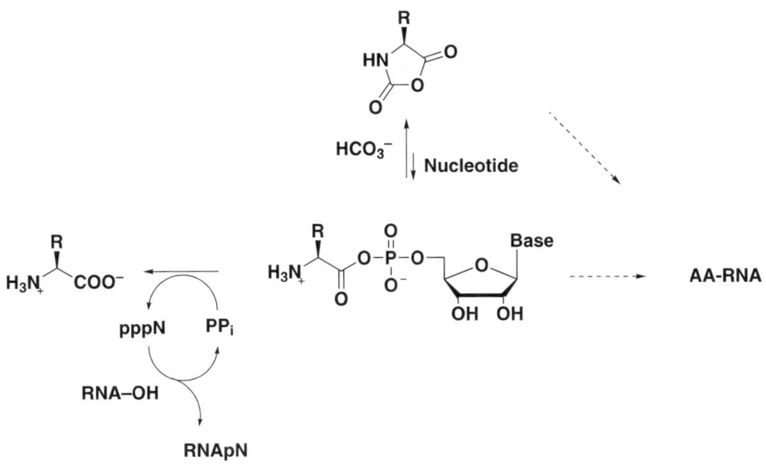

Scheme 36 A plausible process of nucleotide activation using NCAs as energy carriers in an RNA world. For reasons of simplicity, the activation of a separate nucleotide is presented here, but many other alternative systems can be conceived as for example the reaction of a pendant phosphate at the 5′-terminus of an RNA oligomer. The structure of the oligomer may assist the reaction through neighboring group effects and favorable interactions with reactants

form of triphosphates. In this scenario we start from the idea that, because of their availability and their unusually transferable energy content, NCAs played a role of energy carriers capable of activating inorganic phosphate to provide a prebiotic phosphorylation pathway and also capable of activating nucleotides under the form of the mixed anhydrides. Then, phosphate mixed anhydrides with amino acids may have converted nucleotides into nucleoside-5′-triphosphates (**pppN**) by two consecutive phosphorylations. Alternatively, **pppN** formation may have resulted from a reaction with inorganic pyrophosphate that is the reverse of the modern amino acid activation pathway and that needs no energy supply but only needs a catalyst since it is a thermodynamically downhill process.

In this scenario, nucleoside triphosphates may have been polymerized into RNA, regenerating pyrophosphate, which gives rise to a pyrophosphate metabolic cycle as shown in Scheme 39. RNA polymerization would have been dependent on the activity of ribozymes or simpler catalysts [201], which is consistent with the occurrence of an "RNA world" but a world in which *nucleotide and RNA chemistries relied on the energy supplied by amino acid chemistry*. This scenario can be improved by a further or simultaneous stage of peptide synthesis involving aminoacyl-RNA conjugates and the appearance of the translation process because of the selection of individual amino acids by different sequences of RNA. A key step in this archaic process is the formation of amino acid-nucleotide mixed anhydrides from NCAs that could have been improved by one of the first coded peptide catalysts (or by an oligonucleotide moiety) stabilizing the transition state of the reaction with pyrophosphate. But the global efficiency of nucleotide activation through this process can also be increased on the condition that the highly unstable mixed anhydride intermediate is bound and stabilized by the catalyst rendering the equilibrium with NCAs less unfavorable in the presence of CO_2 (stage 2 of Scheme 40). In other words, the efficiency of nucleotide activation could have been the driving force that guided the evolution towards adenylate anhydride binding proteins.

In this scenario, the next key step was the discovery by early living organisms of alternative pathways to synthesize nucleoside triphosphates (stage 3 in Scheme 40), for instance by the introduction of an oxido-reduction metabolism. In the particular case of ATP, the increase in its concentration could have forced the chemical flux of the system to be reverted since the unstable adenylate anhydride was stabilized in the active site of the adenylate binding protein (**E**) allowing the activation of amino acids by ATP through the process that became predominant.

An early emergence of adenylate-forming enzymes can also explain the wide diversification of this family of enzymes [175], many of which are now capable of reacting with carboxylic acids without α-amino group (essential in the hypothetical early mechanism using NCAs as substrates, but then useless) and which is involved in many different biosynthetic pathways.

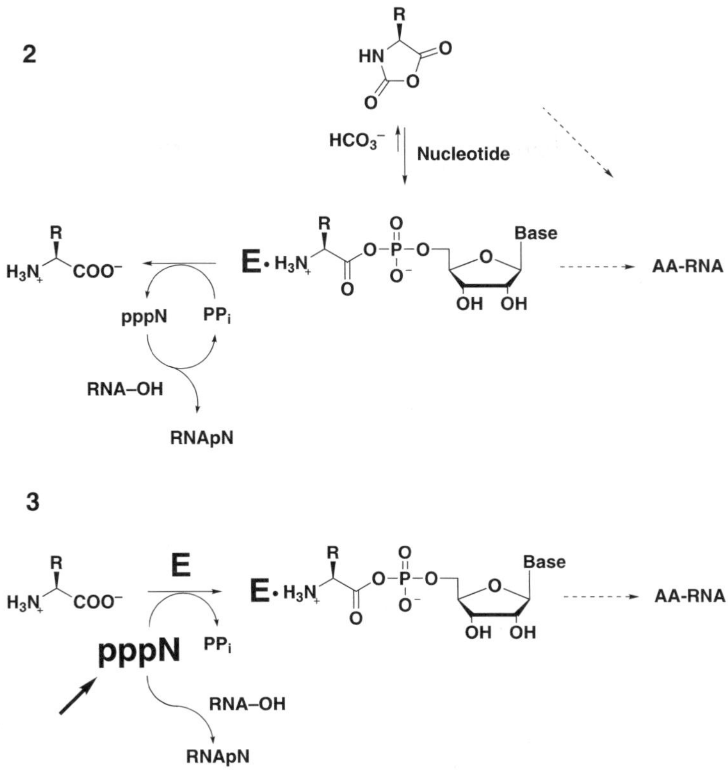

Scheme 37 Second and third stages of the evolution of amino acid activation: (2) The system of Scheme 39 is improved by enzyme catalysis: the mixed anhydride is stabilized by binding making the equilibrium with NCA more favorable; (3) an alternative pathway for nucleoside triphosphate synthesis has been introduced, the chemical flux has been reverted

Aminoacyl phosphate oligonucleotides adaptors [202] were shown to be efficient in peptide bond formation with aminoacylated guide sequences at concentrations much lower than those needed in the reaction of monomers [171]. This rate increase has probably a similar origin [203] as the entropy trap role proposed for the ribosome [204]. We can then conceive systems based on RNA oligomers and capable of the properties mentioned in Scheme 39 with an increased efficiency compared to mononucleotides. Such systems are consistent with the essential role of peptidyl-pre-*t*RNA molecules consisting of a peptidyl moiety bound to an RNA oligomer for the origin of the genetic code [25, 205]. These peptidyl-RNA conjugates may have also served as anchoring devices allowing the RNA part to be concentrated in aggregates formed by the amphiphilic peptide moiety [26, 206]. This idea is consistent with the possibilities that the main utility of the first peptides was not catalysis but simply to hold together polynucleotides [207] and that they formed the primitive apparatus from which evolved both the membrane and the cytoskeleton [208].

4.4 Catalysis by Amino Acids and Peptides

The microenvironment that is found in non-folded polymers can promote some kind of specific "enzyme-like" medium effects (due to desolvation in an hydrophobic environment or to electrostatic interactions) allowing significant increases of the rate of certain reactions that are especially sensitive to medium effects. An efficient process of this kind has been reported using an organic polymer consisting of modified polyethyleneimine [209], with important deductions on the capabilities of primitive enzyme catalysis. A noticeable efficiency and selectivity has been reported for selected reactions of peptide dendrimers [210, 211], in which the sequence and, possibly, the structure are more precisely defined. These kinds of catalytic effects may be related to the reactions that could have been promoted by uncoded prebiotic peptides. However, it must be stressed that high catalytic efficiencies would never be reached by this kind of catalysts in the general case, because a tightly folded and packed structure is needed to fix the geometry of the catalytic site [105], which determines the lower size of efficient polypeptide catalysts to several tenth of residues. Actually, it seems likely that inorganic minerals, inorganic ions, small organic species [44], or nucleotidic cofactors [11] performed most catalytic activities in an abiotic world or at the early stages of biochemical evolution. Interestingly amino acids have been proposed as catalytic cofactors in an RNA world [7]. In this context, several examples of catalysis by amino acids have been reported. Amino acid catalysis, which has also reached an important place in stereoselective organic synthesis [212], has a special importance for the catalysis of carbon-carbon-forming aldol condensations. The catalysis of aldol reactions of dihydroxyacetone by peptides has also been reported [213]. The process is efficient in aqueous solution and, possibly of importance for the prebiotic synthesis of sugars [214, 215] (see above Sect. 4.2).

5 Conclusion

Probably because researchers involved in the study of amino acid and peptide prebiotic chemistry have seen their field of investigation challenged by the possible occurrence of an RNA world, much progress has been made in this field during the last decade. New directions have been explored, leading to prebiotically plausible pathways accounting for both amino acid and peptide production, and peptide chemo-metabolisms have been proposed. Since new indications have been put forward suggesting that life emerged from a co-evolution involving nucleic acid and peptide chemistries, we are confident that further investigations will succeed in proposing scenarios assembling prebiotic building blocks and energy to yield evolved systems. Efficient and selective reactions have been observed for systems involving intramolecular steps, in agreement with the fact that intramolecular catalysis is the only chemical system capable of challenging enzymatic efficiency, which supports its broad involvement in early catalytic systems.

One of the main ideas developed in this chapter has been to emphasize the role played by NCAs, which is supported by a thermodynamic analysis of their unique behavior among activated peptide monomers, but also to suggest that they played a role of energy carriers capable of activating inorganic phosphate and possibly nucleotides. This interaction between NCAs and phosphate or nucleotides is an evolved feature and it would be surprising that it would have been lost by evolution at an intermediate stage before the emergence of the modern protein-nucleic acid world in which this linkage has become prevailing. These ideas are consistent with the possibility of life originating from a peptide-RNA world with a very early emergence of the translation apparatus. They may be supported by further experiments in the next years as well as the role of ribonucleopeptides that could be built using these systems. These directions may lead to overcome the limits of the RNA world hypothesis.

Acknowledgements The authors are grateful to the European COST organization through the D27 action *Prebiotic Chemistry and Early Evolution*, to John Sutherland (University of Manchester, UK) and Franck Selsis (Centre de Recherche Astronomique de Lyon, France) for helpful discussions, to Prof. David Clark (University of Southern Illinois, USA) for making us aware of the free energy content of acetyl adenylate reported in [156], to the CNRS and to the French Ministry of Education and Research for financial support.

Appendix

Assessment of the Free Energy of Hydrolysis of *N*-Carboxyanhydrides

To the best of our knowledge, no assessment of the free energy content of NCAs has been reported. In order to give a basis for the discussion we present the following evaluation built on the assessment of the equilibrium constants corresponding to the different steps of NCA hydrolysis.

The cyclization of carbamic acid **34** into NCA was supposed to correspond to an equilibrium constant that would not be very different from the values reported for the cyclization of succinic acid derivatives **36** ($K_{cyc} = 0.7 \times 10^{-5}$ and 4×10^{-5} for R = H and Me, respectively [216]). The value of $K_{cyc} = 1 \times 10^{-5}$ was therefore selected for carbamic acid **34**.

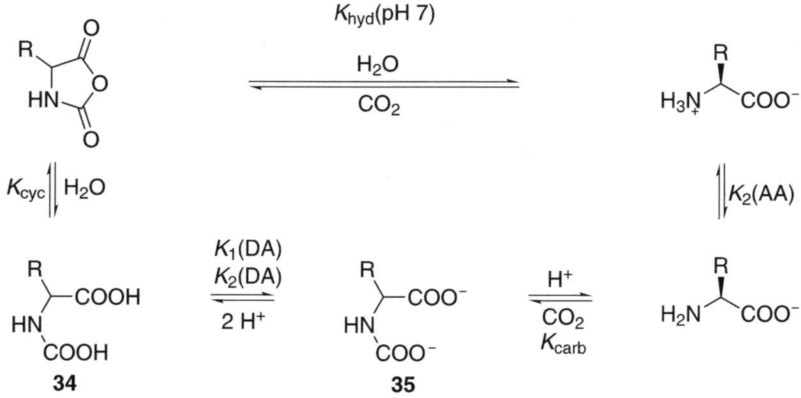

$$K_{hyd}(\mathrm{pH}\ 7) = \frac{K_1(\mathrm{DA})\ K_2(\mathrm{DA})\ (1 + K_1(\mathrm{carbonate}) / [\mathrm{H}^+])}{K_{cyc}\ K_2(\mathrm{AA})\ K_{cyc}}$$

Scheme 38

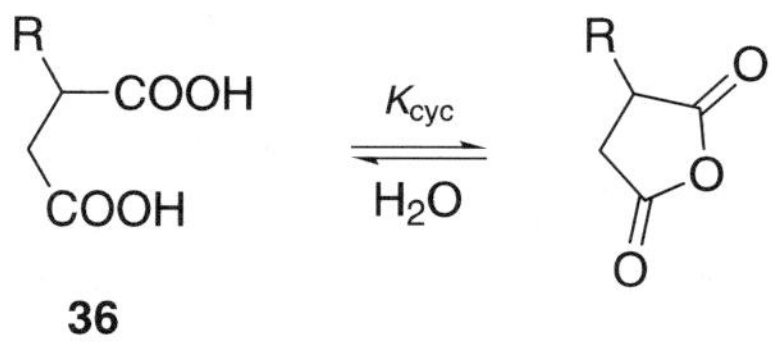

Scheme 39

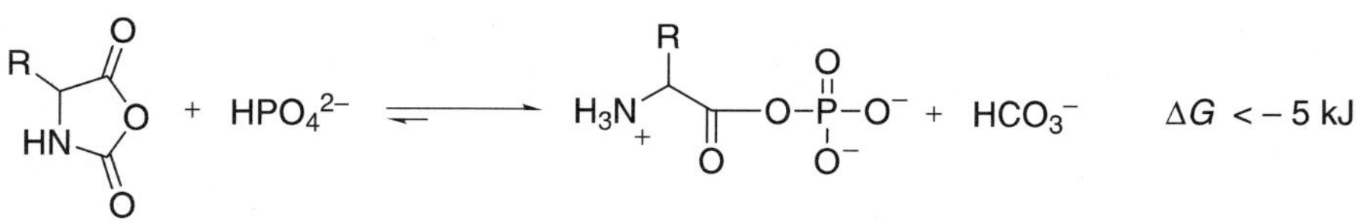

Scheme 40

The pK_A values for the carboxylic and carbamic acid functionalities of diacid **34** were estimated to pK_1(DA) = 3.7 (3.66 and 3.88 for *N*-ethoxycarbonyl-glycine and *N*-carbamoyl-glycine, respectively [154]) and pK_2(DA) = 6.2 (5.25 for carbamic acid and 5.33 for glycyl-glycine carbamate [217] and taking into account a 0.9 pK_A shift due to electrostatic destabilization of the dianion), respectively. A value of pK_2(AA) = 9.8 was selected for the dissociation of the protonated amino acid. The equilibrium constant for the formation of carbamate **35** was estimated to $K_{carb} \sim 1 \times 10^{-5}$from a series of experimental values [217–221].

The overall equilibrium constant K_{hyd}(pH 7) $\sim 5 \times 10^{10}$ was obtained by combining these values at pH 7 and taking into account the fact that CO_2 is mainly under the form of HCO_3^- ion (pK_1 (carbonate) = 6.4), which corresponds to an approximate value $\Delta G^{o\prime} \sim -60$ kJ mol^{-1}.

This estimate is consistent with the fact that the reaction of Val-NCA with phosphate yielding valyl-phosphate is not reversible within the limits of detection allowed by NMR spectroscopy [165]. The free energy of phosphate reaction must therefore be situated below – 5 kJ mol^{-1}, which is in agreement with the estimated value of – 50 kJ mol^{-1} stated in Table 1 for aminoacyl phosphates.

References

1. Miller SL (1953) Science 117:528
2. Bada JL, Lazcano A (2003) Science 300:745
3. Lazcano A, Bada JL (2003) Origins Life Evol Biosphere 33:235
4. Pizzarello S (2004) Origins Life Evol Biosphere 34:25
5. Stubbe J (2000) Curr Opin Struct Biol 10:731
6. Szathmary E (1993) Proc Natl Acad Sci USA 90:9916
7. Szathmary E (1999) Trends Genet 15:223
8. Cech TR (1990) Angew Chem Int Ed 29:759
9. Altman S (1990) Angew Chem Int Ed 29:749
10. Orgel LE (1968) J Mol Biol 38:381–393.
11. White HB (1976) J Mol Evol 7:101
12. Gilbert W (1986) Nature 319:618
13. Gesteland RF, Cech TR, Atkins JF (1999) The RNA World, 2nd edn. Cold Spring Harbor Laboratory Press, New York
14. Orgel LE (2004) Crit Rev Biochem Mol Biol 39:99
15. Orgel LE (1998) Trends Biochem Sci 23:491
16. Noller HF, Hoffarth V, Zimniak L (1992) Science 256:1416
17. Nissen P, Hansen J, Ban N, Moore PB, Steitz TA (2000) Science 289:920
18. Nielsen PE (1993) Origins Life Evol Biosphere 23:323
19. Böhler C, Nielsen PE, Orgel LE (1995) Nature 376:578
20. Nelson KE, Levy M, Miller SL (2000) Proc Natl Acad Sci USA 97:3868
21. Lacey JC, Weber AL, White WE (1975) Origins Life 6:273
22. Borsenberger V, Crowe MA, Lehbauer J, Raftery J, Helliwell M, Bhutia K, Cox T, Sutherland JD (2004) Chem Biodiversity 1:203
23. Smith JM, Borsenberger V, Raftery J, Sutherland JD (2004) Chem Biodiversity 1:1418
24. Tan B, Lee MC, Cui M, Liu T, Chen Z-Z, Li Y-M, Ju Y, Zhao Y-F, Chen K, Jiang H-L (2004) J Mol Struct (Theochem) 672:51
25. Di Giulio M (1997) J Mol Evol 45:571
26. Cavalier-Smith T (2001) J Mol Evol 53:555
27. Weber AL (2002) Origins Life Evol Biosphere 32:333
28. Nicolis G, Prigogine I (1977) Self-Organization in Nonequilibrium Systems. Wiley, New York
29. Martin H (2005) In: Gargaud M, Barbier B, Martin H, Reisse J (eds) Lectures in Astrobiology – Part I: The Early Earth and Other Cosmic Habitats for Life (Advances in Astrobiology and Biogeophysics). Springer, Berlin Heidelberg New York, p 91
30. Kasting JF (1993) Science 259:920
31. Chyba CF, Sagan C (1992) Nature 355:125

32. Bauer SJ (2002) Eur Space Agency, Special Publication SP-518:21
33. Miller SL (1955) J Am Chem Soc 77:2351
34. Strecker A (1850) Liebigs Ann Chem 75:27
35. Schlesinger G, Miller SL (1973) J Am Chem Soc 95:3729
36. Taillades J, Commeyras A (1974) Tetrahedron 30:127
37. Taillades J, Commeyras A (1974) Tetrahedron 30:2493
38. Taillades J, Commeyras A (1974) Tetrahedron 30:3407
39. Béjaud M, Mion L, Commeyras A (1976) Bull Soc Chim Fr:1425
40. Moutou G, Taillades J, Bénefice-Malouet S, Commeyras A, Messina G, Mansani R (1995) J Phys Org Chem 8:721
41. Miller SL, Van Trump JE (1981) In: Wolman Y (ed) Origin of life. Reidel, Dordrecht, p 135
42. Pascal R, Taillades J, Commeyras A (1980) Tetrahedron 36:2999
43. Taillades J, Beuzelin I, Garrel L, Tabacik V, Bied C, Commeyras A (1998) Origins Life Evol Biosphere 28:61
44. Pascal R (2003) Eur J Org Chem:1813
45. Jencks WP (1975) Adv Enzymol Relat Areas Mol Biol 43:219
46. Bergs H (1929) German Patent 566.094
47. Rousset A, Lasperas M, Taillades J, Commeyras A (1980) Tetrahedron 36:2649
48. Taillades J, Rousset A, Lasperas M, Commeyras A (1986) Bull Soc Chim Fr:650
49. Blagoeva IB, Pojarlieff IG, Dimitrof VS (1978) J Chem Soc Perkin Trans II:887
50. Yamagata Y, Mori T (1982) Origins Life 12:41
51. Commeyras A, Boiteau L, Vandenabeele-Trambouze O, Selsis F (2005) In: Gargaud M, Barbier B, Martin H, Reisse J (eds) Lectures in Astrobiology – Part II: From Prebiotic Chemistry to the Origins of Life on Earth (Advances in Astrobiology and Biogeophysics). Springer, Berlin Heidelberg New York, p 501
52. Hennet RJC, Holm NG, Engel MH (1992) Naturwissenschaften 79:361
53. Marshall WL (1994) Geochim Cosmochim Acta 58:2099
54. Alargov DK, Deguchi S, Tsujii K, Horikoshi K (2002) Origins Life Evol Biosphere 32:1
55. Kobayashi K, Kohara M, Dokiya M, Yanagawa H (1997) Viva Origino 25:167
56. Islam MN, Kaneko T, Kobayashi K (2003) Bull Chem Soc Jpn 76:1171
57. Kobayashi K, Kaneko T, Saito T, Oshima T (1998) Origins Life Evol Biosphere 28:155
58. Kobayashi K, Kaneko T, Saito T (1999) Adv Space Res 24:461
59. Kobayashi K, Masuda H, Ushio K-I, Ohashi A, Yamanashi H, Kaneko T, Takahashi J-I, Hosokawa T, Hashimoto H, Saito T (2001) Adv Space Res 27:207
60. Plankensteiner K, Reiner H, Schranz B, Rode BM (2004) Angew Chem Int Ed 43:1886
61. Weber AL (1998) Origins Life Evol Biosphere 28:259
62. Weber AL (2001) Origins Life Evol Biosphere 31:71
63. Cronin JR (1998) In: Brack A (ed) The molecular origins of life. Cambridge University Press, Cambridge, p 119
64. Cronin JR, Chang S (1993) In: Greenberg JM, Mendoza-Gomez CX, Pirronello V (eds) The Chemistry of Life's Origins. Kluwer, Dordrecht, p 209
65. Shimoyama A, Ogasawara R (2002) Origins Life Evol Biosphere 32:165
66. Cooper GW, Cronin JR (1995) Geochim Cosmochim Acta 59:1003
67. Meierhenrich UJ, Munoz Caro GM, Bredehoeft JH, Jessberger EK, Thiemann WHP (2004) Proc Natl Acad Sci USA 101:9182
68. Pizzarello S, Cooper GW (2001) Meteorit Planet Sci 36:897
69. Cronin JR, Pizzarello S (1997) Science 275:951
70. Pizzarello S, Cronin JR (2000) Geochim Cosmochim Acta 64:329
71. Pizzarello S, Zolensky M, Turk KA (2003) Geochim Cosmochim Acta 67:1589

72. Crisma M, Moretto A, Formaggio F, Kaptein B, Broxterman QB, Toniolo C (2004) Angew Chem Int Ed 43:6695–6699
73. Pizzarello S, Weber AL (2004) Science 303:1151
74. Maurette M (1998) Origins Life Evol Biosphere 28:385
75. Brinton KLF, Engrand C, Glavin DP, Bada JL, Maurette M (1998) Origins Life Evol Biosphere 28:413
76. Maurette M, Duprat J, Engrand C, Gounelle M, Kurat G, Matrajt G, Toppani A (2000) Planet Sp Sci 48:1117
77. Snyder LE (1997) Origins Life Evol Biosphere 27:115
78. Sorrell WH (1997) Astrophys Space Sci 253:27
79. Kuan Y-J, Charnley SB, Huang H-C, Tseng W-L, Kisiel Z (2003) Astrophys J 593:848
80. Charnley SB, Ehrenfreund P, Kuan Y-J (2003) Physics World 16:35
81. Walch SP, Bauschlicher CWJ, Ricca A, Bakes ELO (2001) Chem Phys Lett 333:6
82. Mendoza C, Ruette F, Martorell G, Rodriguez LS (2004) Astrophys J 601:L59
83. Takahashi J-I, Hosokawa T, Masuda H, Kaneko T, Kobayashi K, Saito T, Utsumi Y (1999) Appl Phys Lett 74:877
84. Takano Y, Masuda H, Kaneko T, Kobayashi K (2002) Chem Lett:986
85. Munoz Caro GM, Meierhenrich UJ, Schutte WA, Barbier B, Segovia AA, Rosenbauer H, Thiemann WHP, Brack A, Greenberg JM (2002) Nature 416:403
86. Bernstein MP, Dworkin JP, Sandford SA, Cooper GW, Allamandola LJ (2002) Nature 416:401
87. Kobayashi K, Kasamatsu T, Kaneko T, Koike J, Oshima T, Saito T, Yamamoto T, Yanagawa H (1995) Adv Space Res 16:21
88. Kasamatsu T, Kaneko T, Saito T, Kobayashi K (1997) Bull Chem Soc Jpn 70:1021
89. Takano Y, Ushio K, Kaneko T, Kobayashi K, Hashimoto H (2003) Chem Lett 32:612
90. Devienne FM, Barnabé C, Couderc M, Ourisson G (1998) C R Acad Sci Paris, Serie IIc 1:435
91. Devienne FM, Barnabé C, Ourisson G (2002) C R Chimie 5:651
92. Ehrenfreund P, Bernstein MP, Dworkin JP, Sandford SA, Allamandola LJ (2001) Astrophys J 550:L95
93. Schwell M, Dulieu F, Leach S (2001) Eur Space Agency, Special Publication SP-496:133
94. Barbier B, Chabin A, Chaput D, Brack A (1998) Planet Sp Sci 46:391
95. Boillot F, Chabin A, Bure C, Venet M, Belsky A, Bertrand-Urbaniak M, Delmas A, Brack A, Barbier B (2002) Origins Life Evol Biosphere 32:359
96. Barbier B, Coll P, Cottin H, Leach S, Maurel M-C, Raulin F, Tepfer D, Brack A (2002) Eur Space Agency, Special Publication SP-518:59
97. Basiuk VA, Douda J, Navarro-Gonzalez R (1999) Adv Space Res 24:505
98. Basiuk VA, Navarro-Gonzalez R, Basiuk EV (1998) Origins Life Evol Biosphere 28:167
99. Brack A, Barbier B, Bertrand M, Chabin A, Westall F (2002) Eur Space Agency, Special Publication SP-518:435
100. Jakubke H-D, Kuhl P, Könnecke A (1985) Angew Chem Int Ed 24:85
101. Ivanov IP, Todorov NP, Petkov DD (1993) Tetrahedron 49:2307
102. Jencks WP (1976) In: Fasman GD (ed) Handbook of Biochemistry and Molecular Biology, 3rd edn, Vol I Physical and Chemical Data. CRC Press, Cleveland, p 296
103. Smith RM, Hansen DE (1998) J Am Chem Soc 120:8910
104. Radzicka A, Wolfenden R (1996) J Am Chem Soc 118:6105
105. Corey MJ, Corey E (1996) Proc Natl Acad Sci USA 93:11428
106. Lahav N, White D, Chang S (1978) Science 201:67

107. Flegmann AW, Tattersall R (1979) J Mol Evol 12:349
108. Rode BM (1999) Peptides 20:773
109. Kirby AJ (1980) Adv Phys Org Chem 17:183
110. Plankensteiner K, Righi A, Rode BM (2002) Origins Life Evol Biosphere 32:225
111. Bujdak J, Rode BM (1999) Origins Life Evol Biosphere 29:451
112. Bujdak J, Rode BM (1999) J Mol Catal A: Chem 144:129
113. Rode BM, Son HL, Suwannachot Y, Bujdak J (1999) Origins Life Evol Biosphere 29:273
114. Porter TL, Eastman MP, Hagerman ME, Price LB, Shand RF (1998) J Mol Evol 47:373
115. Li Y-M, Yin Y-W, Zhao Y-F (1992) Int J Pept Protein Res 39:375
116. Zhao Y-F, Cao P-S (1994) J Biol Phys 20:283
117. Ju Y, Zhao Y-F, Sha Y-W, Tan B (1995) Phosphorus Sulfur Silicon 101:117
118. Zhou W, Ju Y, Zhao Y-F, Wang Q, Luo G (1996) Origins Life Evol Biosphere 26:547
119. Zhao Y-F, Cao P-S (1996) In: Chela-Flores J, Raulin F (eds) Chemical Evolution: Physics of the Origin and Evolution of Life (Proceedings of the 4th Trieste Conference on Chemical Evolution, Trieste, Sept. 4–8, 1995). Kluwer, Dordrecht, p 279
120. Zhao Y-F, Cao P-S (1999) Pure Appl Chem 71:1163
121. Fu H, Han B, Zhao Y-F, Tu G-Z, Xu L, Lu Q, Wang J-Z, Xiao H-Z (2003) Bioorg Chem 31:122
122. Cheng CM, Liu XH, Li Y-M, Ma Y, Tan B, Wan R, Zhao Y-F (2004) Origins Life Evol Biosphere 34:455
123. Chen Z-Z, Tan B, Li Y-M, Zhao Y-F (2003) J Org Chem 68:4052
124. Kricheldorf HR (1987) α-Aminoacid-N-Carboxy-Anhydrides and Related Heterocycles. Springer, Berlin Heidelberg New York
125. Blair NE, Bonner WA (1980) Origins Life 10:255
126. Ehler KW (1976) J Org Chem 41:3041
127. Brack A (1987) Origins Life 17:367
128. Hessler AM, Lowe DR, Jones RL, Bird DK (2004) Nature 428:736
129. Ferris JP, Hill AR, Liu R, Orgel LE (1996) Nature 381:59
130. Jencks WP (1981) Proc Natl Acad Sci USA 78:4046
131. Page MI, Jencks WP (1971) Proc Natl Acad Sci USA 68:1678
132. Mandolini L (1986) Adv Phys Org Chem 22:1
133. Cacciapaglia R, Di Stefano S, Mandolini L (2004) Acc Chem Res 37:113
134. Wieland T, Jaenicke F (1956) Liebigs Ann Chem 599:125
135. Hay RW, Main L (1968) Aust J Chem 21:155
136. Brack A (1982) Biosystems 15:201
137. Lacey JC, Senaratne N, Mullins DW (1984) Origins Life 15:45
138. Mullins DW, Senaratne N, Lacey JC (1984) Origins Life 14:597
139. Paecht-Horowitz M, Berger J, Katchalsky A (1970) Nature 228:636
140. Armstrong DW, Seguin R, McNeal CJ, Macfarlane RD, Fendler JH (1978) J Am Chem Soc 100:4605
141. Williams A, Douglas KT (1975) Chem Rev 75:627
142. Hegarty AF, Frost LN, Coy JH (1974) J Org Chem 39:1089
143. Collet H, Bied C, Mion L, Taillades J, Commeyras A (1996) Tetrahedron Lett 37:9043
144. Taillades J, Collet H, Garrel L, Beuzelin I, Boiteau L, Choukroun H, Commeyras A (1999) J Mol Evol 48:638
145. Collet H, Garrel L, Commeyras A, unpublished results
146. Commeyras A, Collet H, Boiteau L, Taillades J, Vandenabeele-Trambouze O, Cottet H, Biron J-P, Plasson R, Mion L, Lagrille O, Martin H, Selsis F, Dobrijevic M (2002) Polym Int 51:661

147. Commeyras A, Boiteau L, Vandenabeele-Trambouze O, Selsis F (2005) In: Gargaud M, Barbier B, Martin H, Reisse J (eds) Lectures in Astrobiology – Part II: From Prebiotic Chemistry to the Origins of Life on Earth (Advances in Astrobiology and Biogeophysics). Springer, Berlin Heidelberg New York, p 531
148. Navarro-Gonzalez R, Molina MJ, Molina LT (1998) Geophys Res Lett 25:3123
149. Mather TA, Pyle DM, Allen AG (2004) Geology 32:905
150. Plasson R, Biron J-P, Cottet H, Commeyras A, Taillades J (2002) J Chromatogr A 952:239
151. Liu R, Orgel LE (1997) Nature 389:52
152. Maurel M-C, Orgel LE (2000) Origins Life Evol Biosphere 30:423
153. Leman L, Orgel LE, Ghadiri MR (2004) Science 306:283
154. Kortüm G, Vogel W, Andrussow K (1961) Dissociation constants of organic acids in aqueous solution. Butterworths, London
155. Di Sabato G, Jencks WP (1961) J Am Chem Soc 83:4393
156. Mahler HR, Cordes EH (1966) Biological Chemistry, 1st edn. Harper & Row, New York
157. Holler E, Calvin M (1972) Biochemistry 11:3741
158. Stojanoski K, Zdravkovski Z (1993) J Chem Educ 70:134
159. Paecht-Horowitz M, Katchalsky A (1967) Biochim Biophys Acta 140:14
160. Lewinsohn R, Paecht-Horowitz M, Katchalsky A (1967) Biochim Biophys Acta 140:24
161. Wickramasinghe NSMD, Staves MP, Lacey JC (1991) Biochemistry 30:2768
162. Warden JT, McCullough JJ, Lemmon RM, Calvin M (1974) J Mol Evol 4:189
163. Holler E, Rainey P, Orme A, Bennett EL, Calvin M (1973) Biochemistry 12:1150
164. Berg P (2003) J Biol Chem 278:40417
165. Biron J-P, Pascal R (2004) J Am Chem Soc 126:9198
166. Kluger R, Li X, Loo RW (1996) Can J Chem 74:2395
167. Kluger R, Loo RW, Mazza V (1997) J Am Chem Soc 119:12089
168. Yarus M (1999) Curr Opin Chem Biol 3:260
169. Illangasekare M, Yarus M (1997) J Mol Biol 268:631
170. Kumar RK, Yarus M (2001) Biochemistry 40:6998
171. Tamura K, Schimmel P (2003) Proc Natl Acad Sci USA 100:8666
172. Tamura K, Schimmel P (2004) Science 305:1253
173. De Duve C (1992) In: Tran Thanh Van J, Tran Thanh Van K, Mounolou J-C, Schneider J, McKay C (eds) Frontiers of Life. Editions Frontières, Gif-sur-Yvette p 1
174. Lipmann F (1971) Science 173:875
175. Di Giulio M (1996) Biosystems 39:159
176. Wieland T, Lambert R, Lang HU, Schramm G (1955) Liebigs Ann Chem 597:181
177. Weber AL, Orgel LE (1979) J Mol Evol 13:193
178. Wieland T, Bartmannn W (1956) Chem Ber 89:946
179. Wieland T, Euler KE (1958) Chem Ber 91:2305
180. Wieland T, Franz J, Pfleiderer G (1955) Chem Ber 88:641
181. Lazcano A, Miller SL (1996) Cell 85:793
182. Akabori S (1955) Kagaku 25:54
183. Akabori S (1959) In: Oparin AI, Braunstein AE, Pasinskii AG, Pavlovskaya TE (eds) Proceedings of the first international symposium: The origin of life on the Earth (Moscow, 1957). Pergamon Press, New York, p 189
184. Hanafusa H, Akabori S (1959) Bull Chem Soc Jpn 32:626
185. Ehler KW, Orgel LE (1976) Biochim Biophys Acta 434:233
186. Wen K, Orgel LE (2001) Origins Life Evol Biosphere 31:241
187. Blocher M, Hitz T, Luisi PL (2001) Helv Chim Acta 84:842

188. Hitz T, Luisi PL (2003) Helv Chim Acta 86:1423
189. Hitz T, Luisi PL (2000) Biopolymers 55:381
190. Hill AR, Bohler C, Orgel LE (1998) Origins Life Evol Biosphere 28:235
191. Vaidyanathan R, Kalthod VG, Ngo DP, Manley JM, Lapekas SP (2004) J Org Chem 69:2565
192. Liu R, Orgel LE (1997) J Am Chem Soc 119:4791
193. Liu R, Orgel LE (1998) Origins Life Evol Biosphere 28:47
194. Duvernay F, Chiavassa T, Borget F, Aycard J-P (2004) J Am Chem Soc 126:7772
195. Huber C, Wächtershäuser G (1998) Science 281:670
196. Plasson R, Bersini H, Commeyras A (2004) Proc Natl Acad Sci USA 101:16733
197. Collet H, Boiteau L, Taillades J, Commeyras A (1999) Tetrahedron Lett 40:3355
198. Huber C, Eisenreich W, Hecht S, Wächtershäuser G (2003) Science 301:938
199. Pizzarello S, Weber AL (2004) Science 303:1151
200. Bloch K (1996) Chem Biol 3:405
201. Gao K, Orgel LE (2000) Origins Life Evol Biosphere 30:45
202. Schimmel P, Henderson B (1994) Proc Natl Acad Sci USA 91:11283
203. Tamura K, Alexander RW (2004) Cell Mol Life Sci 61:1317
204. Sievers A, Beringer M, Rodnina MV, Wolfenden R (2004) Proc Natl Acad Sci USA 101:7897
205. Di Giulio M (2003) Origins Life Evol Biosphere 33:479
206. Terenzi S, Biaa E, Nguyen-Trung NQ, Strazewski P (2003) Angew Chem Int Ed 42:2909
207. Orgel LE (1987) Cold Spring Harbor Symp Quant Biol 52:9
208. Hartman H (1992) In: Tran Thanh Van J, Tran Thanh Van K, Mounolou J-C, Schneider J, McKay C (eds) Frontiers of Life. Editions Frontières, Gif-sur-Yvette p 163
209. Hollfelder F, Kirby AJ, Tawfik DS (2001) J Org Chem 66:5866
210. Lagnoux D, Delort E, Douat-Casassus C, Esposito A, Reymond J-L (2004) Chem Eur J 10:1215
211. Clouet A, Darbre T, Reymond J-L (2004) Angew Chem Int Ed 43:4612
212. Jarvo ER, Miller SJ (2002) Tetrahedron 58:2481
213. Tang Z, Yang Z-H, Cun L-F, Gong L-Z, Mi A-Q, Jiang Y-Z (2004) Org Lett 6:2285
214. Wu Y-S, Shao W-Y, Zheng C-Q, Huang Z-L, Cai J, Deng Q-Y (2004) Helv Chim Acta 87:1377
215. Kofoed J, Machuqueiro M, Reymond J-L, Darbre T (2004) Chem Commun:1540
216. Eberson L, Welinder H (1971) J Am Chem Soc 93:5821
217. Johnson SL, Morrison DL (1972) J Am Chem Soc 94:1323
218. Caplow M (1968) J Am Chem Soc 90:6795
219. Morrow JS, Keim P, Gurd FRN (1974) J Biol Chem 249:7484
220. Davis AJ, O'Brien P, Nunn PB (1993) Bioorg Chem 21:309
221. Chen J-G, Sandberg M, Weber SG (1993) J Am Chem Soc 115:7343

Top Curr Chem (2005) 259: 123–165
DOI 10.1007/b137067

Published online: 11 August 2005

Stochastic "Mirror Symmetry Breaking" via Self-Assembly, Reactivity and Amplification of Chirality: Relevance to Abiotic Conditions

Isabelle Weissbuch · Leslie Leiserowitz · Meir Lahav (✉)

Department of Materials and Interfaces, The Weizmann Institute of Science, 76100 Rehovot, Israel
isabelle.weissbuch@weizmann.ac.il, leslie.leiserowitz@weizmann.ac.il, meir.lahav@weizmann.ac.il

1	**Introduction**	124
2	**Stochastic Mirror Symmetry Breaking by Spontaneous Self-Assembly**	126
2.1	Chiral Crystals and "Absolute" Asymmetric Synthesis	127
2.2	Chiral Surfaces of 3-D Crystals	128
2.3	Reduction in Symmetry of Host/Guest Mixed Crystals	129
2.4	Chiral Steps and Kinks on Surfaces of Metallic Crystals	134
2.5	Mirror Symmetry Breaking of 2-D Clusters on Surfaces	134
2.6	Spontaneous Mirror Symmetry Breaking by Formation of Clusters in Bulk-Solutions	136
3	**Amplification of Chirality in Systems Undergoing Spontaneous Mirror Symmetry Breaking**	138
3.1	"Absolute" Asymmetric Synthesis by Crystallization and Topochemical Reactions	138
3.2	Mirror Symmetry Breaking via Autocatalytic Crystallization of the System Glycine/Racemic α-Amino Acids	140
3.3	Kondepudi's Model of Mirror Symmetry Breaking by Crystallization under Stirring	142
4	**Non-Linear Effects in Asymmetric Catalysis**	144
4.1	Soai's Auto-Catalytic System for Spontaneous Mirror Symmetry Breaking	145
5	**Generation of Homochiral Biopolymers from Racemates**	147
5.1	Homochiral Polymers via 2-D Self-Assembly and Lattice-Controlled Polymerization	149
5.2	Enantiopure Oligopeptides from Non-Racemic Precursors	150
5.3	Homochiral Oligopeptides in a Phospholipid Environment	153
5.4	Enantiopure Homochiral Oligopeptides Generated by Topochemical Reactions in 3-D Crystals	154
6	**Self-Replication of Biopolymers**	157
7	**Conclusions**	158
	References	159

Abstract Theories of prebiotic life suggest that homochirality emerged in Nature in abiotic times via deterministic or chance scenarios. This chapter deals with experiments demonstrating the feasibility of stochastic mirror symmetry breaking that occurs via autocatalytic processes involving the self-assembly of molecular clusters, 2-D and 3-D crystals, supramolecular organo-metallic catalysts, and polymeric helices and sheets. Once generated spontaneously by chance, chirality can be preserved and propagated to the environment provided that the symmetry breaking step is coupled with a sequential step of efficient amplification via self-replication reactions. Common features for the systems of relevance are that they take into consideration small fluctuations from the racemic state, and they display non-linear kinetic effects induced by diastereoisomeric supramolecular self-assemblies that exhibit different physical or chemical properties.

Keywords Asymmetric autocatalysis · Biomolecular handedness · Chiral clusters · Chiral surfaces · Enantiomorphous crystals · Homochiral peptides · Mirror symmetry breaking · Non-linear kinetics · Self-replication of peptides

1
Introduction

One of the remarkable hallmarks in Nature is the bias observed in biopolymers composed from homochiral L-α-amino acids and D-sugars toward a single handedness. This has resulted in a common perception that the presence of molecules of a single handedness is a unique signature of living systems. Theories for the origin of a single chirality in the biological world fall into two major categories: biotic and abiotic. The first category suggests that selection of one of the enantiomers took place at a late stage in the biological evolution of living matter [1, 2]. The second scenario proposes that chiral materials were formed prior to the appearance of the earlier biopolymers [3]. This surmise is based on experimental work [4], independently supported by theoretical considerations [5], where directed complementary oligomerisations in templates of nucleotides performed on enantiomerically-pure primers of nucleic acids were strongly inhibited in the presence of racemic ribonucleotides. For this reason, it has been a challenge to demonstrate that spontaneous mirror symmetry breaking of racemic mixtures is feasible under conditions similar to those occurring in primordial times.

Deterministic and stochastic processes should be taken into account. The deterministic scenario invokes transfer of the intrinsic chirality of the universe to the biopolymers of life [6, 7]. Ab initio theoretical estimates that take into account the chiral electroweak forces indicate that the L-amino acids and the D-sugars are more stable than their corresponding enantiomers [8, 9]. The minute energy differences between these enantiomeric pairs, under Darwinian reaction kinetics in a flow reactor, were invoked to account for the biomolecular handedness that arose when life began [7]. Several reports describing deterministic mechanisms that could have induced mirror symmetry breaking are experimentally untenable or have been disproved [10, 11]. A second deterministic effect that might have played a role in this context of mirror symmetry breaking is the modus operandi of circularly polarized light (CPL), or the combination of plane polarized light and an oriented magnetic field [12, 13]. Both asymmetric synthesis and the induced enantioselective photo-resolution of racemates of amino acids and other organic molecules with high enantiomeric excesses (*ee*) have been observed [14]. The role played by CPL in the origin of molecular chirality has attracted interest, since astrophysicists have observed that light emitted from interstellar stars is circularly polarized and thus might have been instrumental in the generation of extra-terrestrial enantiopure materials that landed on Earth [15]. Also relevant in this context are the reports that amino acids extracted from the Murchison meteorite contain some non-natural amino acids enriched with the L-enantiomers [16, 17]. Although the origin of the handedness of these amino acids has not been fully determined, one assumes that they have been formed by enantioselective photodecomposition of the racemic mixtures, achieved by irradiation with CPL [18, 19]. Two recent studies have demonstrated that these non-natural amino acids might have served as chiral auxiliary molecules for the preparation of non-racemic sugars [20] and peptides of natural amino acids that assume 3_{10} helical structures [21].

A stochastic surmise suggests that spontaneous mirror symmetry breaking might have occurred in systems that were far from equilibrium and that had undergone phase transitions [22]. In these transformations, at variance with the deterministic ones, there is equal probability of obtaining either the D- or the L-enantiomer in excess. On the other hand, such stochastic routes are theoretically well understood and experimentally-feasible in the laboratory, and therefore some of them might be realistic under prebiotic conditions. In order to preserve, as well as to propagate, the handedness once generated by stochastic scenarios, it is imperative to couple the first event of mirror symmetry breaking with a sequential step of efficient autocatalytic amplification. A number of such successful experiments had been reported over the years, some of which are presented below.

Several comprehensive reviews had appeared over the years on the topic of the origin of homochirality [14, 23–29]. Here, we summarize several recently-reported experimental studies that are pertinent to the problem of stochastic mirror symmetry breaking, with an emphasis on experiments performed in the authors' laboratory. In particular, symmetric systems undergoing spontaneous segregation into supramolecular chiral assemblies are considered, with the focus on systems where the formed chiral architectures self-replicate auto-catalytically and propagate their handedness to the rest of the environment. These architectures comprise chiral crystalline clusters and nuclei of organic and inorganic crystals, organo-metallic complexes, liquid crystals, growing and dissolving surfaces, and linear polymers as template primers. Reactions in chiral self-assemblies can provide simple mechanistic routes for the generation of homochiral biopolymers from racemic or non-racemic activated monomers of low enantiomeric imbalance.

2 Stochastic Mirror Symmetry Breaking by Spontaneous Self-Assembly

The models of Prigogine and Kondepudi [22, 30] consider non-chiral systems that undergo a phase transition. The models assume that at the singular point of the transition, new dissipative architectures might be formed. If these structures are chiral, the system bifurcates into two new heterochiral phases. Where a single or a small number of dissipative assemblies are formed, they are dominated by species of one handedness. At this early stage of the transition, the mirror symmetry of the system has been broken spontaneously. However, during the formation of an additional large number of dissipative structures, they will be of both handednesses and the emerging phase will be symmetric. Symmetric systems can be driven towards chirality provided that the first "Adam" [31] chiral species formed stochastically after the phase transition generates new fresh assemblies of the same handedness via a autocatalytic mechanism, or that it inhibits the formation of new ones of opposite handedness via enantiomeric cross-inhibition. It has been demonstrated that both effects operate in unison and that the handedness of the new phase is determined by the handedness of the first "Adam" assembly.

Two additional models that encompass similar stochastic features have been proposed; the mathematical model by Frank on the polymerization reactions of amino acids [32], and the autocatalytic scheme by Calvin [33], Scheme 1.

Several systems are pertinent for the realization of these models, including crystalline arrays and lattice-controled reactivity therein, organization of

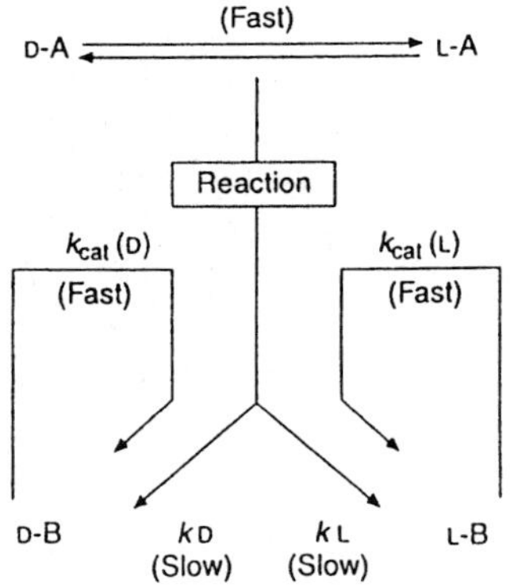

Scheme 1

molecules on surfaces, asymmetric autocatalysis, and asymmetric induction in the formation of isotactic polymers.

2.1 Chiral Crystals and "Absolute" Asymmetric Synthesis

Non-chiral or chiral molecules undergoing fast racemisation that self-assemble in enantiomorphous two-dimensional (2-D) or three-dimensional (3-D) crystalline domains reside in a chiral environment within these supramolecular architectures. In such systems, the overall composition of the crystallites is either non-racemic or even of a single handedness, depending upon the conditions of the crystallization. Homochiral crystals can be obtained when the crystallization starts from a single nucleus, but their handedness cannot not be predicted in the absence of chiral auxiliaries, and the enantiomeric excess of the crystallites will vary from one experiment to another. The term "total asymmetric transformation" has been coined to discern such processes. Early examples of inorganic molecules displaying such an effect comprise crystals of quartz, sodium chlorate, and sodium bromate. In the 1950s, this phenomenon was also made manifest in organic crystals by the pioneering study of Havinga [34] on methyl-ethyl-allyl-anilinium iodide, followed by experiments on the inclusion complexes of tri-*o*-thymotide [35]. Non-chiral crystals such as potassium dichromate and boric acid have been shown to self-assemble in helical mesoscopic morphologies via diffusion-limited growth in various gel matrixes, without the assistance of chiral auxiliary molecules [36, 37].

Very recently, these processes were extended for the generation of helical morphologies of K_2SO_4 crystals [38] generated in the presence of polyacrylic acid, Fig. 1a,b, and for the preparation of helices from the achiral $BaCO_3$ nanocrystals [39, 40] grown in the presence of racemic hydrophilic block copolymers.

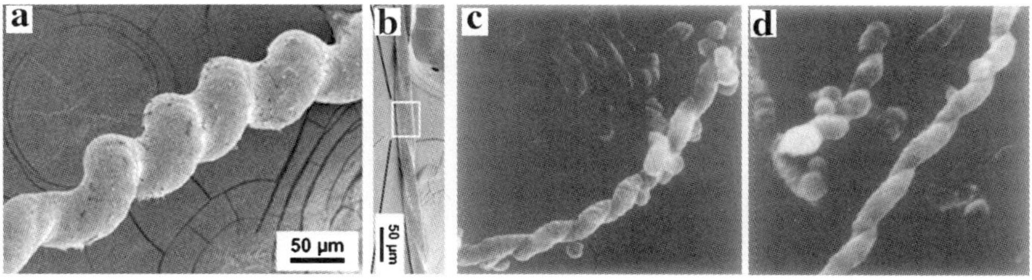

Fig. 1 **a, b** Helical morphologies and backbone obtained by crystallization of K_2SO_4 with an increase in polyacrylic acid concentration. (Reproduction from [38]. Copyright 2005, American Chemical Society) **c, d** Left- and right-handed helical silica (diameter 100–110 nm) used for "absolute" asymmetric catalysis (Reproduced from [56]. Copyright 2005, Elsevier)

Chiral crystals generated from non-chiral molecules have served as reactants for the performance of so-called "absolute" asymmetric synthesis. The chiral environments of such crystals exert asymmetric induction in photochemical, thermal and heterogeneous reactions [41]. Early reports on successful "absolute" asymmetric synthesis include the γ-ray-induced isotactic polymerization of *trans-trans*-1,3-pentadiene in an all-*trans* perhydrophenylene crystal by Farina et al. [42] and the gas-solid asymmetric bromination of p,p'-dimethyl chalcone, yielding the chiral dibromo compound, by Penzien and Schmidt [43]. These studies were followed by the $2\pi + 2\pi$ photodimerization reactions of non-chiral dienes, resulting in the formation of chiral cyclobutanes [44–48]. In recent years more than a dozen such syntheses have been reported. They include unimolecular di-π-methane rearrangements and the Nourish Type II photoreactions [49] of an achiral oxo- [50] and a thio-amide [51] into optically active β-lactams, photo-isomerization of alkyl-cobalt complexes [52], asymmetric synthesis of two-component molecular crystals composed from achiral molecules [53] and, more recently, the conversion of non-chiral aldehydes into homochiral alcohols [54, 55].

Right- and left-handed helical silica, Fig. 1c,d, has been successfully used as an auxiliary in a catalytic reaction used in "absolute" asymmetric synthesis [56].

2.2 Chiral Surfaces of 3-D Crystals

Crystals interact with molecules of the environment via the surfaces that delineate them. Consequently, several of their properties, such as their morphology, structure and symmetry of solid-solutions and their etch-pit patterns formed upon partial dissolution, depend on an interplay between the surface structures of the crystal faces and the composition of the solution. For example, crystallization of a racemate undergoing spontaneous resolution in the

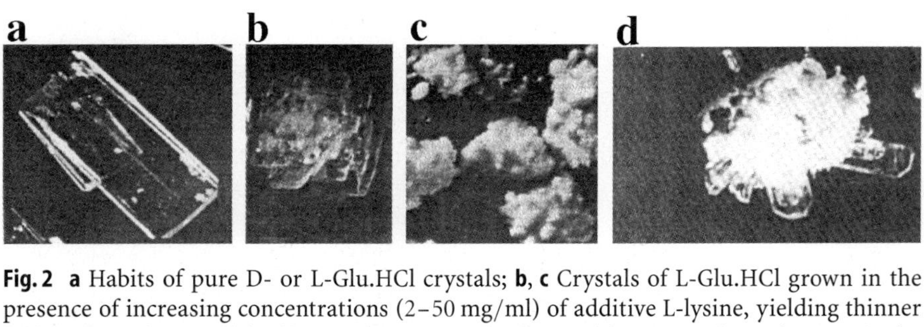

Fig. 2 **a** Habits of pure D- or L-Glu.HCl crystals; **b, c** Crystals of L-Glu.HCl grown in the presence of increasing concentrations (2–50 mg/ml) of additive L-lysine, yielding thinner and thinner plates and finally powder. **d** Crystals obtained from DL-Glu.HCl grown in the presence of additive L-lysine. The powder is the L-enantiomorph, whereas the unaffected crystals are D-Glu.HCl

presence of "tailor-made" chiral additives, which are adsorbed enantiospecifically at the surface of one of the crystals, results in the precipitation of the two enantiomorphs displaying very different morphologies [57]. This approach provides an interesting modification of the classical Pasteur method on the separation of enantiomorphous crystals by entrainment. For example, the enantioselective habit changes observed for the enantiomorphous crystals of L-glutamic acid monohydrochloride (Glu.HCl) grown in the presence of L-lysine are shown in Fig. 2a–c. Crystallization of DL-Glu.HCl in the presence of L-lysine yielded L-crystals as a powder covering the unaffected D-crystals, (Fig. 2d).

2.3
Reduction in Symmetry of Host/Guest Mixed Crystals

Mechanistic studies on crystal growth of mixed host/guest crystals have demonstrated that their formation is kinetically-driven. The first step, which must take place prior to occlusion of the guest molecules into the bulk of the host crystal, is their recognition by structured sites present at the various faces of the host crystal. If the symmetry of the crystal face through which the guest molecules are occluded is lower than that of the crystal, the mixed crystal undergoes a reduction in symmetry as compared to the host. This phenomenon has been demonstrated by X-ray and neutron diffraction studies in crystals of D-asparagine monohydrate grown in the presence of D-aspartic acid [58].

Centrosymmetric crystals are delineated by pairs of chiral faces of opposite handedness that may interact enantioselectively in solution with guest molecules tailored so that a small fraction thereof may be adsorbed and eventually occluded into the crystal bulk through these chiral faces. Following this mechanism, many non-chiral host crystals were "converted", on growth, into a conglomerate of chiral sectors of the mixed crystal. The handedness of each sector is predetermined by the chirality of the face through which the guest molecules have been occluded [59–61]. This concept is illustrated with an "absolute" asymmetric synthesis performed in the centrosymmetric host crystals of *E*-cinnamamide, $Ph-CH:CH-CONH_2$, with non-chiral guest molecules of *E*-cinnamic acid, $Ph-CH:CH-COOH$ [60]. According to the above mechanism, the cinnamic acid guest molecules are occluded enantioselectively at chiral sites of one handedness through the $+b$ pole of the host crystal (Fig. 3a) and at enantiomeric sites through the $-b$ pole, resulting in the transformation of the centrosymmetric host into a mixed crystal composed of two chiral halves that are coherently compounded. The change in morphology from the pure host crystal to the host/guest mixed crystal, that exhibits well-developed $\{011\}$ surfaces through which the guest cinnamic acid molecules have been occluded, is shown in Fig. 3b,c. Irradiation of the crystal sectors at the $+b$ and $-b$ poles with UV light yielded, in

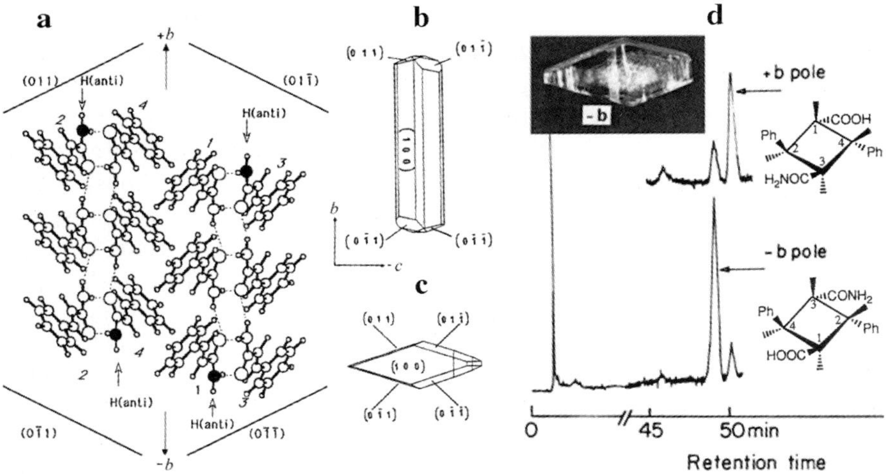

Fig. 3 **a** Packing arrangement of *E*-cinnamamide crystal viewed along the *a*-axis. The {011} faces and the four symmetry-related sites in black circles are denoted; **b** Morphology of pure crystal grown from methanol/water; **c** Morphology of mixed crystal grown in the presence of *E*-cinnamic acid; **d** Photograph of the mixed crystal with the assigned – *b* pole and enantiomeric analysis by gas chromatography of derivatives of the photodimerization products isolated from the + *b* and – *b* poles of the specimen crystal

addition to the centrosymmetric α-truxilic acid, enantiomerically-enriched mixed acid/amide cyclobutane products, as shown in Fig. 3d.

An "absolute" asymmetric synthesis has been also successfully performed via the heterogeneous addition of OsO_4 on the double bond of achiral tiglic acid that crystallizes in a centrosymmetric crystal where one of the enantiotopic faces is blocked during the reaction [62].

The mechanism of reduction in crystal symmetry has been demonstrated for various sectors of centrosymmetric host crystals of *E*-cinamamide grown in the presence of a *E*-2-thienyl-acrylamide guest [59] and by second harmonic generation measurements in mixed crystals of a centrosymmetric host nitro-amine Schiff base grown in the presence of the corresponding dinitro-molecules [63, 64]. Reduction in symmetry was also observed in mixed inorganic crystals of isomorphous salts of $Ba(NO_3)_2$ and $Pb(NO_3)_2$ by polarized light microscopy [65]. Using this method, it has been demonstrated that the surfaces of di(11-bromoundecanoyl)peroxide crystals can discriminate between equally-sized bromine and methyl groups of van der Waals radii of 1.9 and 2.0 Å [66, 67].

The occurrence of reduction of symmetry is of particular importance in the mirror symmetry breaking process of racemic α-amino acids, accomplished with the assistance of crystals of glycine grown at interfaces. When grown from aqueous solutions, glycine crystallizes in its centrosymmetric α-polymorph (space group $P2_1/n$). This crystal is composed from chiral layers of glycine molecules (colored red and in green in Fig. 4a). The non-chiral molecules of glycine in solution become chiral in the crystalline environment because they assume a chiral orientation.

When glycine crystals are grown in aqueous solutions in the presence of racemic mixtures of amino acids, they display a plate-like morphology with two well-expressed chiral enantiotopic (010) and (0-10) faces, Fig. 4b. During growth, the D-amino acids interact enantioselectively with the (010) face by virtue of their α-amino acid moieties and thus replace the 'red' glycine host molecules so that their side chains emerge from the crystal surface and thus do not interfere with the intra-layer binding process. A minor fraction of these amino acid guest molecules would be occluded within the bulk of the glycine crystal on growth. By symmetry, the L-amino acids would be occluded into the bulk of the glycine crystal through the (0-10) face. As a result of this process, racemic α-amino acids can undergo segregation into enantiomers upon occlusion within glycine crystals, Fig. 5.

Furthermore, if the glycine crystals are grown at an interface that blocks the growth at one of the enantiotopic faces, say (010), then only the L-enantiomer of the racemic α-amino acids will be occluded within the crystals through their (0-10) face being exposed to solution, thus "converting" the non-chiral host glycine crystal into a homochiral mixed crystal of single handedness. This transformation can be illustrated with glycine crystals grown in the presence of N^{ε}-(2,4-dinitrophenyl)-L-lysine. Crystals that exposed only their (010) face to solution during growth had not occluded the

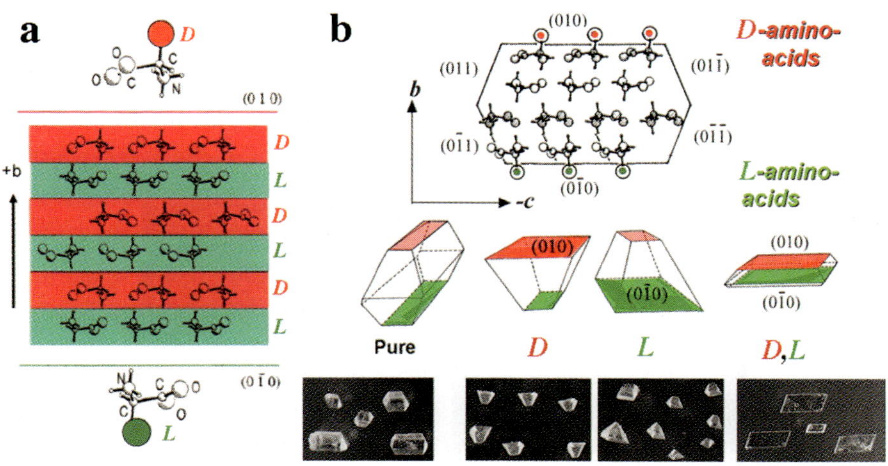

Fig. 4 **a** Packing arrangement of α-glycine with the crystal heterochiral layers of chiral glycine molecules colored in *red* and *green*; **b** Morphology and photographs of the crystals of pure α-glycine as well as those grown in the presence of D-, L-, and DL-α-amino acid additives

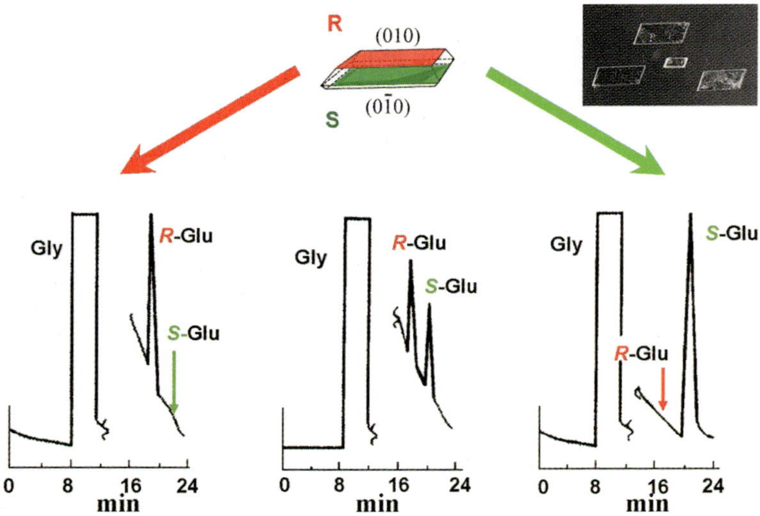

Fig. 5 Enantiomeric HPLC analyses of the occluded additives in plate-like crystals of glycine grown in the presence of DL-glutamic acid; (*left* to *right*) sample taken from the + *b* pole, sample from a whole crystal and sample taken from the – *b* pole

yellow dye and are therefore white, whereas the crystals exposing the (0-10) face to the solution during growth are yellow, Fig. 6.

One may envisage that such conglomerates of crusts of glycine crystals might be spread to yield enantio-enriched environments, as in the mechanism proposed by Welch [68].

Diastereoisomeric interactions between chiral surfaces of non-chiral crystals and chiral molecules present in solution are demonstrated by the formation of etch pits. Etch pits were only formed on the (010) face of an α-glycine crystal partially dissolved in an undersaturated solution containing D-alanine, whereas the (0-10) face does not exhibit etch pits, Fig. 7a [69].

Fig. 6 Photographs of white crusts and yellow crusts of glycine crystals grown at the air/aqueous solution interface in the presence of N^{ε}-(2,4-dinitrophenyl)-L-lysine and leucine in ratios of $L/D > 1$ and $L/D < 1$, respectively. The white crystals exposed their (010) faces towards the solution whereas the yellow crystals exposed their (0-10) faces

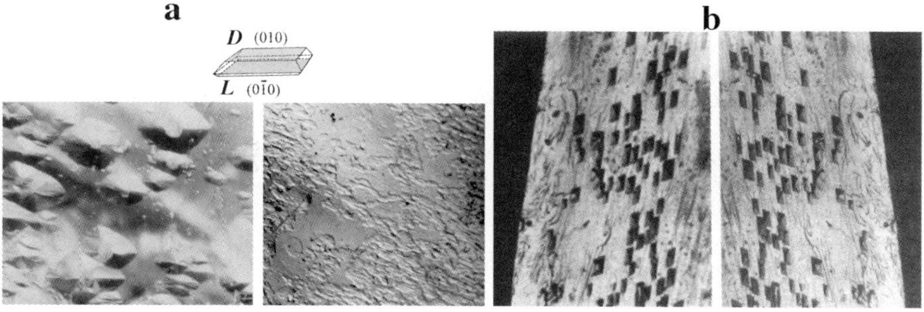

Fig. 7 **a** Photographs of the (010) and (0-10) faces of plate-like α-glycine crystals after etching in the presence of D-alanine; **b** The (010) and (0-10) faces of a cleaved α-glycine crystal subsequently etched in the presence of DL-alanine

When crystals of α-glycine were cleaved at the {010} plane, exposing (010) and (0-10) surfaces that were subsequently etched in a solution containing *DL*-alanine, they revealed mirror symmetry-related etching patterns, as clearly seen in Fig. 7b.

Similar mirror symmetry breaking has been reported for example in the growth of crystals of glycyl-glycine with racemic mixtures of glycyl-leucine, Fig. 8 [70].

Hazen et al. [71, 72] showed that aspartic acid is absorbed chiroselectively from a racemic mixture at chiral steps present at faces of non-chiral calcite

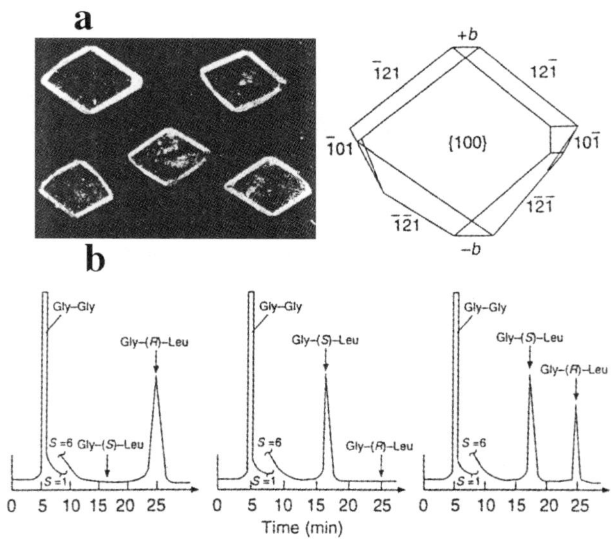

Fig. 8 Glycyl-glycine crystals grown in the presence of DL-glycyl-leucine. **a** Photographs and morphology; **b** Enantiomeric HPLC analyses of samples taken from single crystals cut at the + *b* and – *b* poles and sample from the whole crystal (*left* to *right*)

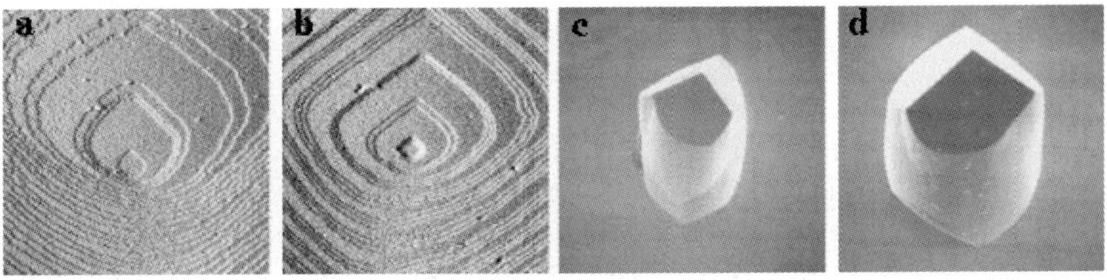

Fig. 9 **a**, **b** AFM images showing the effect of amino acids on calcite growth-hillocks following addition of supersaturated solutions with L- and D-aspartic acid, respectively. **c**, **d** Morphology of calcite crystals grown in the presence of L- and D-aspartic acid, respectively (Reproduced from [74]. Copyright 2001, Nature)

crystals. Cody and Cody [73] have reported enantioselective effects induced by amino acids on the chiral surfaces of gypsum crystals. More recently, Orme et al. [74] reported that calcite crystals develop chiral growth hillocks when grown in the presence of L- or D-aspartic acid, Fig. 9.

Experiments of this kind suggest that similarly kinked chiral sites present at surfaces of minerals might operate as catalytic centers for asymmetric synthesis.

2.4 Chiral Steps and Kinks on Surfaces of Metallic Crystals

Gellman et al. have postulated that symmetric surfaces of single crystals of metals possessing kinked steps are inherently chiral [75]. Low electron emission diffraction (LEED) studies on Pt and Cu faces of single crystals had demonstrated that the steps at these faces are indeed enantiomeric. As such, these surfaces assume enantiospecific catalytic properties. The chiral response to the presence of such chiral kinks at surfaces has been demonstrated in a number of systems, as in the enantioselective electrooxidation of racemic glucose using single crystals of Pt as electrodes that expose a given kinked face to the solution [76]. Similarly, an enantioselective desorption of chiral alcohols and methyl cyclohexanone has been shown to take place from non-chiral kinked faces of the Cu crystal [77].

2.5 Mirror Symmetry Breaking of 2-D Clusters on Surfaces

Two-dimensional (2-D) crystallites are generally of a lower symmetry than 3-D crystals. The molecules in the 2-D crystallites cannot pack across a center of inversion as they most commonly do in 3-D. By applying modern analytical tools such as scanning tunneling and probe microscopy (STM, SPM), transmission electron microscopy (TEM), grazing incidence X-ray diffraction (GIXD), electrospray ionization (ESI) and matrix-assisted laser-desorption ionization time-of-flight (MALDI-TOF) mass-spectrometry, and optical methods such as epifluorescence, Brewster angle spectroscopy, circular dichroism, it became possible to characterize these clusters by assigning their structures at the molecular level in some examples [78, 79]. Studies using STM [80–82] have demonstrated that racemates, when deposited on graphite, gold, copper or on other supports, undergo spontaneous resolution in two dimensions. This effect is illustrated here with the self-assembly of non-chiral molecules of 1-nitro-naphthalene [83, 84]. Although these molecules are non-chiral in solution or in the gas phase, when deposited on a gold surface, they form a library of clusters of different sizes. The dominant decamers of this mixture were imaged using STM and shown to assume a chiral morphology, Fig. 10. According to these measurements (supported by calculations) each cluster is composed from eight molecules of one handedness and two molecules of opposite chirality. These enantiomorphous *L* and *R* clusters were physically separated with the assistance of a STM tip.

Similar 2-D enantiomorphous domains were obtained from the non-chiral nucleic acid base adenine deposited on copper [85] and on MoS_2 surface [86, 87] and the deposition of cysteine on gold [88]. Evidence for strong chiral preference in interactions of nucleic acid bases and amino acids has been shown for the self-assembly of phenylglycine molecules on gold surfaces on which adenine molecules had been previously deposited [89].

Enantiospecific addition of substituted ethylenes containing double bonds to Si dimers at silicon surfaces to form Si – C bonds via $2\pi + 2\pi$ addition reactions has been demonstrated by applying direct scanning probe microscopic measurements [90].

Early studies by surface-pressure-area ($\pi - A$) isotherms provided indirect evidence for the spontaneous resolution of some racemates at the air/water interface [91, 92]. These studies were followed by monolayer imaging using epifluorescence [93–95] and Brewster angle microscopy [96–100].

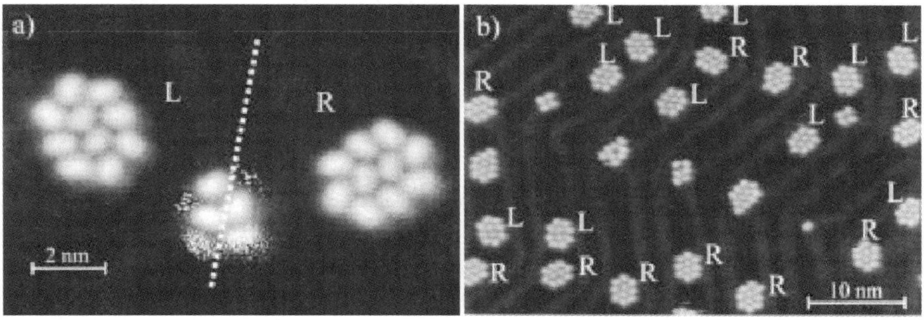

Fig. 10 **a** STM images of two-dimensional chiral decamers (denoted *L* and *R*) formed by 1-nitronaphthalene molecules on the Au (111) surface at 50 K; **b** Surface with 0.1 monolayer coverage exhibits about 85% of enantiomorphous decameric clusters (Reproduced from [83]. Copyright 1999, Wiley)

Recently, the GIXD method has been successfully applied in order to determine the packing arrangements in 2-D crystallites of racemic amphiphiles self-assembled on the suface of water [101]. Such racemates can form either racemic 2-D crystallites, where the two enantiomers are related via glide symmetry, or they undergo a spontaneous segregation into enantiomorphous crystalline domains. Enantiomeric disorder is possible in both types of 2-D crystallites. The GIXD studies have demonstrated that long-chain (C_{11}–C_{17}) α-amino acids as well as γ-stearyl-glutamic acid self-assemble on the surface of water into racemic 2-D crystallites, as opposed to N^{ε}-alkanoyl(C_{12}–C_{22})-lysine amphiphiles and N^{α}-myristoyl-alanine, that undergo spontaneous segregation into enantiomorphous 2-D crystallites [102–104].

The formation of chiral 2-D crystalline domains of non-chiral Cd-arachidate on water has been demonstrated by GIXD and X-ray reflectivity studies [105, 106] and by AFM for Ca-arachidate after transfer of the film onto solid support using the Langmuir-Blodgett technique [107].

More recently, Liu et al. reported that a variety of non-chiral amphiphilic diacetylenes, non-chiral barbituric acids or amphiphilic aryl-benzimidazoles self-assemble into chiral clusters at the air/water interface or on aqueous solutions containing Ag^{+} ions, as demonstrated by CD measurements [108–111]. The chiral macroscopic conformational morphology of the polymers generated from copper salts of non-chiral monomers was imaged after their transfer onto solid support [112, 113].

2.6 Spontaneous Mirror Symmetry Breaking by Formation of Clusters in Bulk-Solutions

Chiral supramolecular architectures generated from non-chiral monomeric units were reported in a number of systems in solution and have been summarized in a comprehensive review [114]. An amplification of chirality on hydrogen-bonded assemblies, controlled by different substitutions and structural variations in the building blocks, has been recently described by Reinhoudt et al. [115].

J.-M. Lehn et al. designed helical architectures via the pre-programming of molecular self-assembly through specific non-bonding interactions [116, 117]. The formation of chiral fiber-like architectures had been also observed from non-chiral monomers such as in the gels of bis-urea building blocks [118].

J-aggregates were demonstrated to undergo "total asymmetric transformations" in solution such that the non-chiral molecules convert into chiral fiber-like associates [119].

A remarkable example of the self-assembly of achiral diprotonated *meso*-tetraphenylsulfonato porphyrins in aqueous solution, yielding chiral homo-associates, was reported by Ribo et al. [120]. These experiments drew great interest since one could determine the sense of chirality of the macroscopic aggregates just by selecting the direction of the vortex by either stirring or rotary evaporation. The diprotonated porphyrins are zwiterionic molecules bearing two positive charges within the porphyrin moiety and negative charges at the two sulfonatophenyl groups located at the *meso*-positions of the porphyrin. The aggregates self-assemble through electrostatic and hydrogen bonds. A 90° folding of porphyrin association produces *P* or *M* chirality due to the small angle of 15–20° between the plane of the porphyrin and the chain alignment, Fig. 11 [121]. The morphology of such chiral self-aggregates imaged by AFM measurements is shown in Fig. 12 [122].

More recently, a related example of the generation of chiral films by spin coating hydrogen-bonded dendritic zinc porphyrin *J*-aggregates, where either one of the two enantiomeric forms of the film is selected by the spinning direction, was reported [123]. Self-aggregation of mixtures of achiral porphyrins bearing oppositely-charged groups was reported to result in achiral 1 : 1 complexes. However, the symmetry of these clusters could be broken by the addition of 10^{-3} M of L- or D-phenylalanine [124].

Chiral octameric aggregates of serine, as formed from enantiopure or non-racemic solutions and detected by ESI mass spectrometry, were reported by a number of laboratories. It has also been demonstrated that these clusters

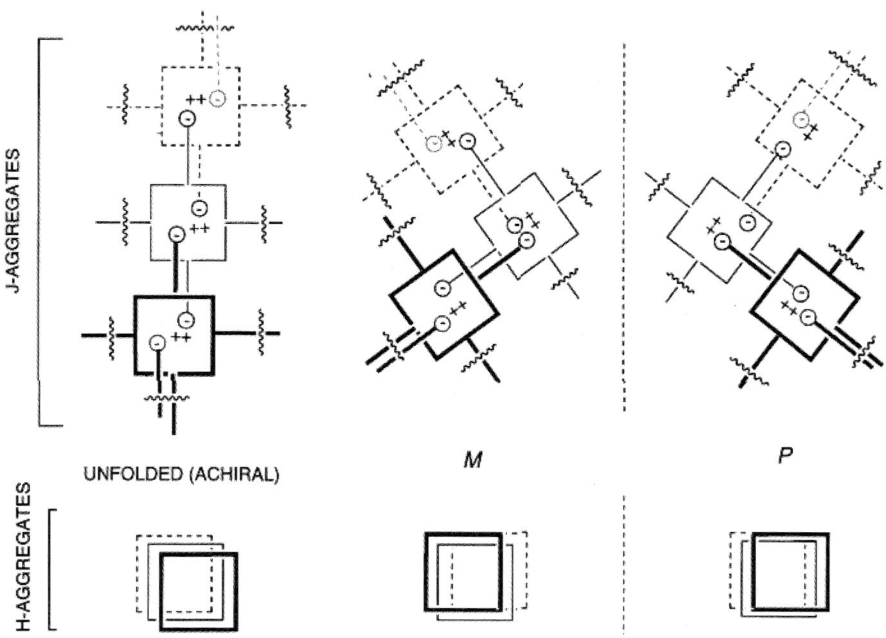

Fig. 11 Schematic porphyrin association at 180° and 90°. The latter shows *P*, *M* chirality due to the angle between the chain alignment and the porphyrin plane (Reproduced from [121]. Copyright 2001, Wiley)

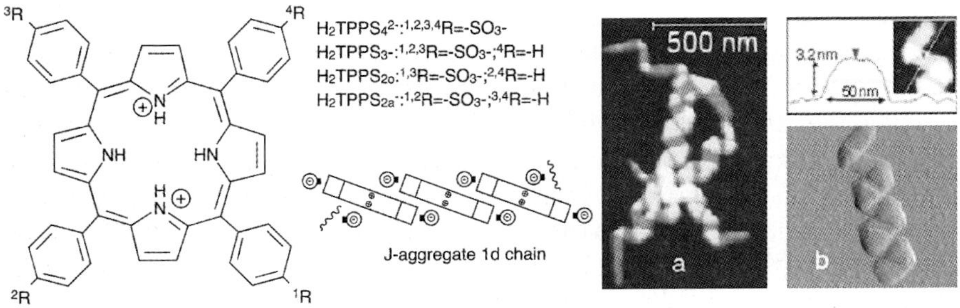

Fig. 12 Molecular formulae and AFM topography and amplitude signal images of helicoidal ribbons of $H_2TPPS_3^-$ aggregates prepared by rotary evaporation of solutions (Reproduced from [122]. Copyright 2003, Royal Society of Chemistry)

incorporate other amino acids of the same chirality in the gas phase. These clusters also react in the vapor phase with glyceraldehydes, glucose, phosphoric acids and other metals to yield chiral products [125, 126].The formation of larger clusters has been observed in more recent studies [127]; however, the chirality of these serine clusters may vary as a function of their size [128].

3 Amplification of Chirality in Systems Undergoing Spontaneous Mirror Symmetry Breaking

The above examples demonstrate that mirror symmetry breaking by self-assembly of non-chiral molecules into chiral architectures is indeed a feasible process. However, in order to preserve the handedness and amplify the stochastically-generated chirality, it is imperative to couple such chance events with efficient sequential autocatalytic processes. We refer now to several experimental systems that illustrate the occurrence of such scenarios. We shall allude in particular to systems undergoing amplification via non-linear asymmetric catalysis processes, via the formation of 2-D and 3-D crystalline systems and amplification of homochiral bio-like polymers in general and oligopeptides in particular.

3.1 "Absolute" Asymmetric Synthesis by Crystallization and Topochemical Reactions

Let us imagine a scenario where a chiral product of a given handedness has been formed by an absolute asymmetric synthesis. Episodic changes in temperature would induce melting of the system that comprises the chiral reactant product. Supplying additional substrate material and then reducing the temperature should result in additional crystallization of the reactant, but this time within a chiral environment. Where the chiral product exerts an asymmetric induction on the nucleation and on the crystal growth processes such that it induces a preferential precipitation of crystals of the same handedness as the Adam crystal, the chirality of these crystals will be auto-catalytically amplified. Several attempts had been made to design such experimental systems, some of which are described below. Green and Heller [129] attempted to probe such a model system in the solid/gas asymmetric bromination of single crystals of p,p'-dimethyl-chalcone. Although they observed an asymmetric effect in a preferred crystallization induced by the chiral dibromide formed via the solid-state reaction, the fresh crystals were of opposite handedness to the parent crystals where the chiral dibromides were generated, Scheme 2.

Similar observations were made during the attempt to amplify chirality in the formation of homochiral cyclobutane polymers via "absolute" $2\pi + 2\pi$ photo-polymerization reactions starting with non-chiral dienes [45–47], as illustrated schematically in Scheme 3.

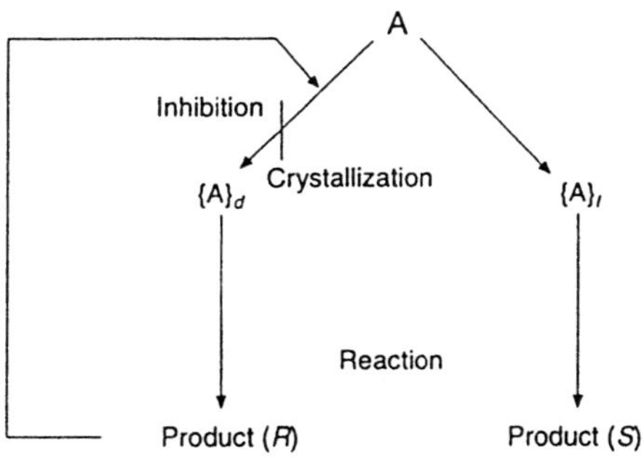

Scheme 2

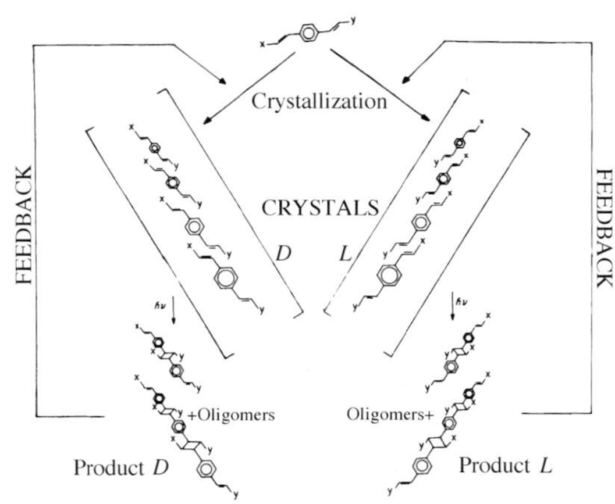

Scheme 3

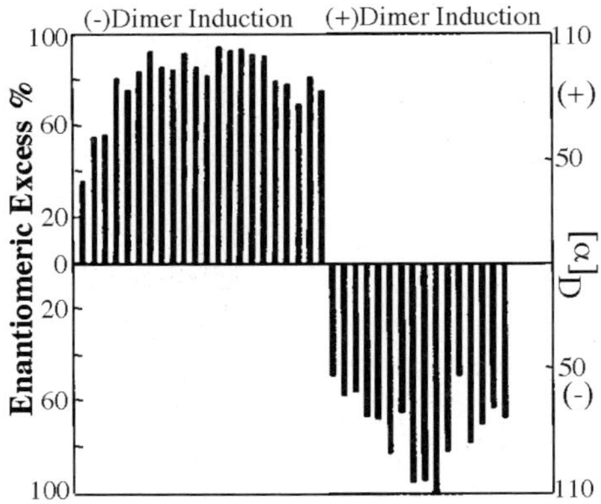

Scheme 4

A number of non-chiral dienes, bearing two different double bonds, were designed by crystal engineering to crystallize into enantiomorphous crystals, where the different double bonds were in the close proximity needed for topochemical photo-polymerization yielding oligo-cyclobutane chiral products. Crystallization experiments with these dienes performed in the presence of optically-resolved dimers demonstrated that crystal nucleation and crystal growth processes were strongly induced. However, it was also noticed that the chiral product operates as an inhibitor of crystal nucleation and growth of the parent enantiomorph rather than as a promoter, as illustrated in the histogram of Scheme 4 [130].

Although these experiments did not provide the desired systems needed to amplify chirality, they were helpful in elucidating the stereochemical mechanism of the role played by additives in the early stages of crystal nucleation. This information was instrumental to the elaboration of appropriate model systems for the amplification of chirality, such as the generation of homochiral lysine via crystals of nickel/caprolactam [131] and the autocatalytic process of the spontaneous segregation of racemic enantiomers of amino acids in aqueous solutions assisted by centrosymmetric glycine crystals grown at interfaces.

3.2
Mirror Symmetry Breaking via Autocatalytic Crystallization of the System Glycine/Racemic α-Amino Acids

In Sect. 2.3 we showed that when glycine crystals are grown at the air/aqueous solution interface in the presence of DL-α-amino acids, only one of its enantiotopic faces, e.g. (010), is exposed to the solution and so it picks up (together with glycine) only the D-α-amino acids, thus converting the centrosymmetric host glycine into chiral mixed crystals. By symmetry, crystals exposing their (0-10) face towards the solution occlude only the L-enantiomers, Scheme 5.

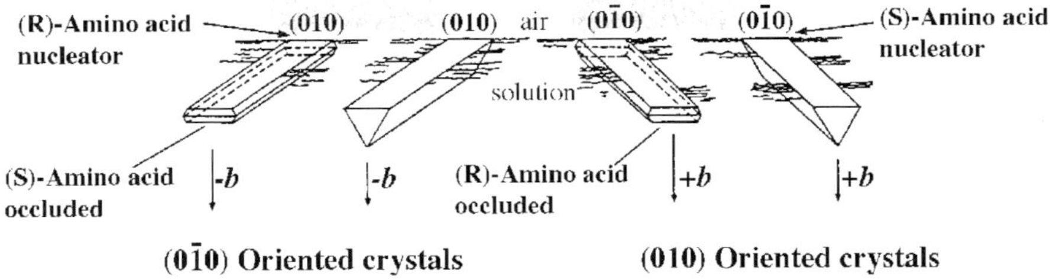

Scheme 5

If by chance a single or a small number of glycine crystals grow at the interface, the solution is enriched with α-amino acids of one handedness. Preservation and proliferation of the handedness generated stochastically by the "Adam" crystals necessitate that the new glycine crystals grown at a latter stage at the air/water interface should adopt the same orientation. Two different effects exert an asymmetric induction in the orientation of the glycine crystals grown at the air/water interface, as required for further amplification of chirality. If the solution contains hydrophobic α-amino acids such as leucine (leu), phenylalanine, and so on, these molecules tend to accumulate at the interface to form 2-D domains that operate as templates for oriented crystallization of the glycine crystals. L-hydrophobic α-amino acids induce crystallization of floating glycine crystals that expose their (010) face toward the solution and occlude only the D-α-amino acids. This asymmetric induction was proven experimentally by the oriented crystallization of glycine crystals induced by the presence of 1% L- or D-leu [132].

Direct evidence for the formation of 2-D clusters of water-soluble, hydrophobic α-amino acids at the air/water interface was provided by the packing arrangement of their water-insoluble counterparts bearing long hydrocarbon chains self-assembled on the water surface, as determined by GIXD. The polar glycine head groups of these amphiphiles form a 2-D hydrogen-bonded net that mimics that of a layer in the glycine crystal, thus serving as a template for the oriented growth of these crystals.

The second effect, that is kinetic in nature, comprises an enantioselective inhibition of the glycine nuclei by the α-amino acids present in the solution. The effect was proven experimentally by achieving complete orientation of the floating glycine crystals when grown in the presence of DL-leucine and hydrophilic L-α-amino acids. The role played by DL-leu was to ensure the nucleation of floating glycine crystals exposing either the (010) or the (0-10) face towards the solution, whereas increasing the concentration of the hydrophilic L-amino acids inhibited the glycine nuclei exposing the (0-10) face towards the solution, thus preventing their further growth. Such experiments yielded floating crystals occluding only the D-leu, leading to an enantiomeric enrichment of L-leu in the solution.

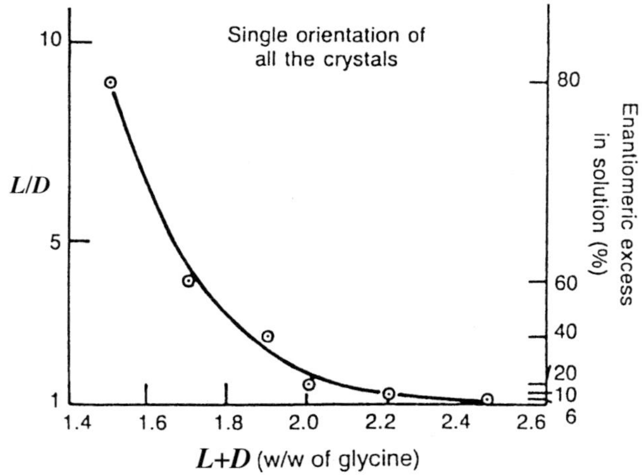

Fig. 13 Correlation between the initial L/D ratio and the total leucine concentration needed for complete orientation of the floating crystals of α-glycine exposing their (010) face towards the solution

Combined operation of the hydrophobic and kinetic effects using non-mixtures of L/D-leu of 53 : 47 (6% *ee*) in a total concentration of 2.4% wt/wt of glycine resulted in the formation of a crust of floating glycine crystals containing only D-leu, thus enriching the initial L-leu *ee* of the solution, Fig. 13 [133–135].

The overall process of mirror symmetry breaking and amplification is illustrated in Scheme 6.

Glycine crystallizes in three different polymorphs. The polymorphs β and γ appear in enantiomorphous space groups. Recent studies have demonstrated that each enantiomorph of β-glycine forms mixed crystals with other α-amino acids of a single handedness. This property has been used in the design of an autocatalytic process, related to that of the α-form, but taking place in bulk aqueous solution [136]. On the other hand, the thermodynamically stable polymorph γ that crystallizes in a $P3_1$ space group cannot be used as a matrix for the mirror symmetry breaking since it does not distinguish between the enantiomeric amino acids [137].

3.3
Kondepudi's Model of Mirror Symmetry Breaking by Crystallization under Stirring

Crystallization processes comprise two sequential steps: crystal nucleation followed by crystal growth. Kondepudi et al. demonstrated in a series of experiments that spontaneous symmetry breaking may be induced by growing crystals of non-chiral molecules such as sodium chlorate, binaphthyl, and p,p'-dimethyl-chalcone, which crystallize as enantiomorphous crystals of

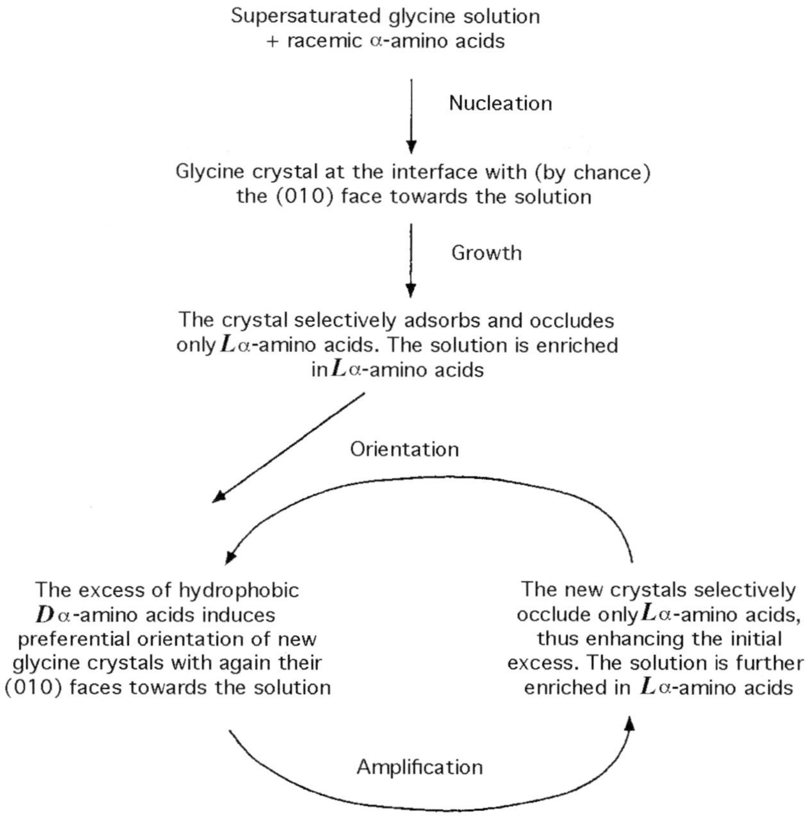

Scheme 6

a single handedness when performed under a regime of stirring with a mechanical stirrer. Since the nucleation step of these crystals is a slow process that requires a high degree of supersaturation, only a small number of crystals are formed in the early stages. On the other hand, the sequential step of crystal growth occurs by secondary nucleation, which requires much a lower degree of saturation. In the absence of stirring, neither handedness dominates among the formed crystals. On the other hand, when the same solution is stirred with a magnetic stirrer, one observes that crystals of a single handedness dominate, as represented in the histograms of Fig. 14.

McBride and Carter [139] have videotaped the collisions associated with these crystallizations, and they observed that once the first crystal was formed the stirring bar produces secondary nuclei. These nuclei, of the same handedness as the "Adam" crystal, are formed in overwhelming numbers and dispersed through the entire solution, serving as seeds for the formation of fresh crystals of the same handedness.

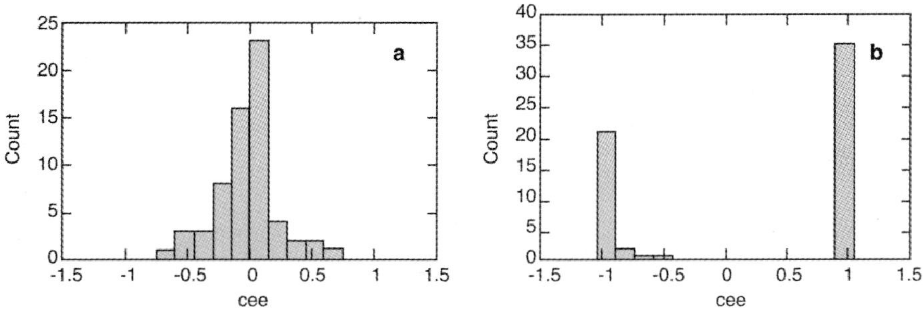

Fig. 14 Histograms of the crystal enantiomeric excess {cee = $(N_L - N_D)/(N_L + N_D)$} of L-crystals for spontaneous chiral symmetry breaking in $NaClO_3$ crystallization: **a** 63 independent unstirred crystallizations; **b** 60 independent stirred crystallizations (Reproduced from [22]. Copyright 2001, American Chemical Society)

Autocatalytic generation of crystals of single handedness alone is not sufficient unless proliferation of the enantiomorphic crystals is not prevented, as proposed in the Frank model for autocatalytic mirror symmetry breaking. In the systems presented above, the appearance of the second class of crystals is prevented since, once the process of crystal growth proceeds, the solution is depleted from the solute, thus reducing the degree of supersaturation to below the one required for additional homogeneous nucleation of fresh crystals. This process has been theoretically modeled by considering a simple autocatalytic reaction scheme combined with chaotic mixing [140, 141]. In the experiments on mirror symmetry breaking by crystallization under stirring conditions, in contrast to those reported by Ribo [120] on self-associates of porphyrins, the handedness of the crystals cannot be controlled by the handedness of the vortex.

Viedma et al. [142] reported experiments that are not in agreement with the explanation that the crystallization experiments under stirring conditions are initiated by an "Adam" crystal but rather via a mechanism that comprises a library of nuclei of both handedness. This proposed mechanism was supported by experiments demonstrating that, in a dissolution-crystallization process, a large symmetric population of D- and L-crystals of sodium chlorate could be driven into crystals of a single handedness via a non-linear autocatalytic-recycling process [143].

4 Non-Linear Effects in Asymmetric Catalysis

The formation of diastereoisomeric aggregates via interactions of chiral molecules of non-racemic mixtures has been successfully applied to the amplification of chirality. Wynberg and Feringa [144] reported that enantiomerically-enriched alkali metal alkoxides that aggregate in solution, where the aggregates operate as chiral catalysts for their own formation from achiral reactants, yield products of the same handedness as the starting material via an auto-catalytic process.

Kagan et al. [145] quantified the results of these non-linear effects by applying theoretical models and providing several additional experimental systems. Furthermore, the origin of the non-linear effects has been also supported by theoretical and calorimetric kinetic studies [146–149].

Efficient amplification of chirality by non-linear effects is achieved in non-racemic systems where the catalysts are organo-metallic reagents. In systems where the metal (M) binds to two organic chiral ligands of opposite handedness (L and D), three diastereoisomeric complexes are formed in a steady state. A fast exchange between the ligands takes place in the solution. The *meso*-complex L-M-D is formed predominantly in cases where its stability is higher than those of the two enantiomeric D-M-D and L-M-L complexes. Accordingly, in non-racemic systems where, say, L > D, one anticipates the formation of two main complexes L-M-L and D-M-L. Furthermore, since these complexes are of different structures, their catalytic properties may differ substantially. In cases where only the homochiral L-M-L complex is an efficient catalyst, the *ee* of the product is much larger than that of the initial ligands, Scheme 7. A comprehensive summary of nonlinear stereochemical effects in asymmetric reactions is presented in [150].

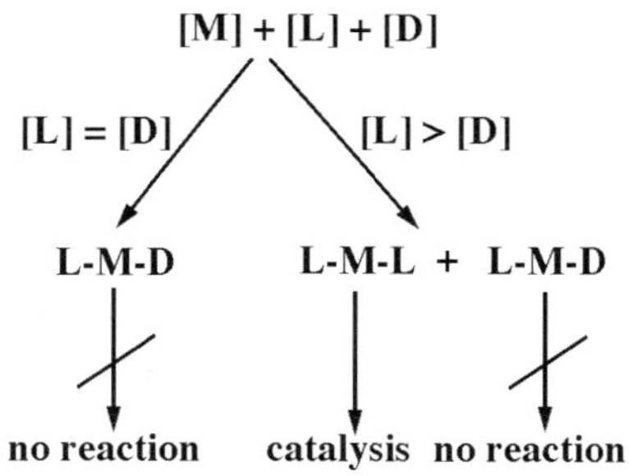

Scheme 7

4.1 Soai's Auto-Catalytic System for Spontaneous Mirror Symmetry Breaking

Soai et al. reported an elegant system where a chiral organo-zinc catalyst of very low *ee* operates as an alkylating agent that converts non-chiral aldehydes into chiral alcohols. This system is related to the complexes displaying non-linear kinetics mentioned above, with one exception: that the chiral product in this reaction serves as a chiral ligand that associates with the organo-Zn catalyst, thus increasing, as the reaction progresses, the concentration of the chiral catalyst. For example, when alcohol molecules of a 10^{-5}% *ee* (approximately 50.000025 : 49.999975) are used as ligands of the Zn catalyst, a non-chiral aldehyde is converted, in a small number of cycles, into a chiral alcohol of 99.5% *ee* [153]. Although there is no easy way to tell the difference between an asymmetric autocatalytic reaction initiated by the tiny *ee* due to the random chance of variations at the level of 10^{12} molecules/mole or one initiated by minuscule quantities of unidentified chiral impurities, the feasibility of amplifying a stochastic fluctuation in a racemate into an enantiopure product has been reported. Indeed, Soai et al. reported the spontaneous asymmetric synthesis of pyrimidyl alcohol, starting from a non-chiral system comprising the alkylation of the carbaldehyde precursor, using diisopropylzinc as the catalyst. In 18 experiments they obtained the formation of D- and in 19 experiments the L-enantiomer [154, 155]. Very recently, Soai et al. [156] reported the asymmetric synthesis of near enantiopure (> 99.5% *ee*) alcohol **1** by asymmetric photodecomposition of the corresponding racemic alcohol **1** by circularly polarized light followed by asymmetric autocatalysis of the non-chiral aldehyde **2**, as shown in Scheme 8.

Singleton and Vo [157] demonstrated that an excess of 60 000 D molecules of the alcohol in the starting solution yields an induced asymmetric induction of the same alcohol after several catalytic cycles. On the other hand, when they performed the same reaction with racemates 54 times, the reaction was driven towards the formation of the D- 27 times and towards the formation of the L-alcohol 27 times. These reference experiments suggest that the reaction is not affected by unforeseen chiral impurities present in the environment. Based on these results, the authors suggested that their observations are more consistent with an asymmetric synthesis originating from the chance *ee* present in the racemic mixture.

Furthermore, Soai et al. demonstrated that the same amplification reactions could also be performed by starting with racemic mixtures of the reactant in the presence of small quantities of chiral crystals of quartz or

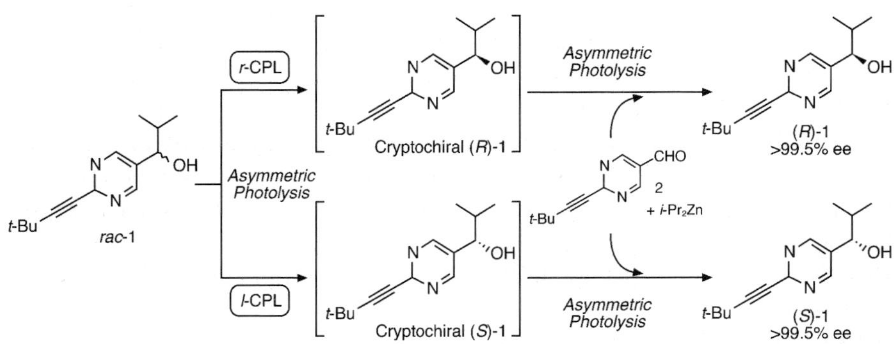

Scheme 8 (Reproduction from [156]. Copyright 2005, ACS)

$NaClO_3$ [158, 159]. The asymmetric induction in these reactions is most probably due to specific interactions between the Zn atom and the chiral surface of the crystals. This amplification reaction could be also performed on racemic mixtures of amino acids or helicenes irradiated first with circularly polarized light to yield non-racemic mixtures of very low *ee* that transformed the solution into a non-racemic mixture, which was then further amplified to an *ee* of beyond 95% [160]. Although the reaction system is prebiotically unrealistic, since it was performed in non-aqueous solutions, it still demonstrates the feasibility of spontaneously driving non-chiral systems towards single handedness by autocatalysis.

5
Generation of Homochiral Biopolymers from Racemates

The above examples suggest that there are various abiotic routes for the production of chiral materials, including amino acids, via mirror symmetry breaking scenarios; however, the *ee* of these chiral products would presumably have been low, due to either their mode of formation or as a result of racemization that took place under the realistic conditions of high energy radiation in prebiotic times. Polymerization reactions of monomers of more than one component performed under ideal conditions obey binomial kinetics, resulting in the formation of a complex mixture of diastereoisomeric polypeptides of composition 2^n where 2 stands for the two enantiomers of the racemate and n is the number of diastereoisomers. Under this regime of reactivity, the distribution of the D and L units is random, resulting in the formation of peptide chains with random sequences. Consequently, the formation of long isotactic chains requires polymerization reactions that occur at conditions different from those in ideal solutions. Apart from several reports that have described the formation of short glycine oligomers [161, 162], non-activated α-amino acids are not regarded as realistic prebiotic precursors for the formation of long peptides. De Duve [163] has proposed thio-esters as possible activated intermediates that might have been formed near volcanic regions [164–166]. Another intermediate is the *N*-carboxyanhydride (NCA) of α-amino acids that can be generated from thio-esters [167] or by other feasible prebiotic routes proposed by the Montpellier group [168, 169].

A number of model reaction schemes have been suggested to overcome the conundrum of the formation of long homochiral oligopeptides from racemates. Wald's model considers a two-step process in the polymerization reactions of activated amino acids [2]. At early stages of the reaction, one anticipates the formation of a random distribution of oligopeptides. Once oligopeptides of homochiral sequences eight units or longer are formed, they self-assemble into helical secondary structures that should exert a very efficient asymmetric induction on latter steps of the propagation reaction by picking up enantiomers of the same handedness. This surmise was experimentally supported by various groups [170, 171] that performed polymerization reactions of NCA amino acids. Spach demonstrated that the helical conformation exerts a superior induction of the polymerization than the chiral terminus of the polypeptide chain. Brack and Spach [172] have proposed that β-sheet secondary structure has even a greater induction effect than the α-helix. The formation of peptide chains assuming either α-helix or β-sheet structures requires at least eight repeating units of the same handedness. The probability of the formation of such chains in ideal solutions is $1/2^8$ i.e. the formation of one such molecule in 256. However, the polycondensation of these monomers occurs at conditions that depart from ideality. Several studies have demonstrated that the dipeptides and the tripeptides formed by the polymerization of racemic protected amino acids are diastereoselective, displaying biases in favor of enantioselective growth [173]. More recent mass spectrometric studies by Luisi et al. on the polymerization of racemates of deuterium enantiolabeled NCA amino acids such as tryptophane, leucine and isoleucine carried out in aqueous solution have demonstrated a departure from a binomial distribution in favor of oligopeptides with homochiral sequences [174, 175]. They also reported that quartz efficiently enhances the mole fraction of homochiral oligo-leucines (for example the 7-mers by a factor of 17) due to selective adsorption of the more stereoregular oligopeptides from an aqueous solution [176].

Enantiomeric enrichment of homochiral peptides was achieved during partial hydrolysis of polypeptides composed from polymers that contain blocks of both random and α-helix [177] or random and β-sheet sequences [178]. Hydrolysis of the atactic part of the polypeptides was found to occur much faster than within the isotactic blocks of α-helix and β-sheets. Based on these observations, Bonner suggested that environmental dry/wet cycles on the primitive Earth might have caused repeated polymerization/hydrolysis cycles that permitted the eventual evolution of optical purity from a small abiotic *ee* value for the amino acids [177, 179].

Magnification of chirality might be achieved in polymers that form helical structures. Green et al. [180] have demonstrated that polyalkyl isocyanates such as polyhexyl isocyanate form left- and right-handed helices, although the monomers themselves are achiral. The cooperative stereochemistry of the side groups is so large that incorporation of as little as 2% of chiral units is sufficient to form copolymers that self-assemble into chiral helices, the handedness of the latter being dictated by the chiral groups. The reaction is so sensitive that even an enantioselective isotopic substitution of the side group results in a preference for one helical form.

In a related system Wittung et al. [181] reported chiral amplification involving achiral polymers composed from glycine repeating units to which a cytosine nucleobase has been attached. These molecules can form complementary base-paired helical duplexes that are analogous to those of DNA and

RNA but of both handednesses. Upon appending a single homochiral residue such as L-lysine at the carboxy terminus of the helix, the "peptide nucleic acid" polymers predominantly fold into duplexes of a single handedness. This phenomenal positive cooperativity can be considered a form of molecular amplification. Additional examples on spontaneous and induced formations of helical morphologies were reviewed [114, 182].

Supramolecular architectures are highly sensitive to chiral perturbations in general, and in systems that form liquid crystals in particular. Small amounts of enantiopure guest molecule added to a nematic host can induce a transition to a cholesteric phase, and the helical organization in the mesoscopic system is very sensitive to the structure of the guest molecule. Chiral amplification was successfully achieved in such liquid crystals, using CPL as the chiral trigger for the phase transition [183].

A similar effect has been reported in the crystallization of non-chiral molecules, where the presence of small amounts of chiral additive forces the entire system to crystallize in an enantiomorphous crystal, which upon further solid-state reaction can be converted into polymers of a single handedness [184, 185]. Chiral auxiliaries, which affect crystal nucleation enantioselectively, have been successfully used for the large-scale optical resolution of enantiomers [186–188].

5.1 Homochiral Polymers via 2-D Self-Assembly and Lattice-Controlled Polymerization

An alternative route for the generation of enantiopure oligopeptides has been elucidated recently by our group. The method comprises the self-assembly of racemic or non-racemic thio-esters or N-carboxyanhydrides of α-amino acids into either 2-D or 3-D crystalline architectures followed by lattice-controlled reactions.

Various comprehensive studies on the polymerization of enantiopure and racemic esters of α-amino acids performed at the air/water interface to yield peptides have been reported over the years [189, 190]. Recent reinvestigations of the products of these reactions by MALDI-TOF MS have demonstrated, however, that they are not longer than dipeptides [191]. For this reason, such esters cannot be regarded as realistic prebiotic model systems for the formation of long oligopeptides. On the other hand, amphiphilic N^{α}-carboxyanhydrides [192] and thio-esters [193] of α-amino acids yield longer oligopeptides.

GIXD studies have demonstrated that racemic N^{ε}-alkanoyl-lysines and their corresponding N^{α}-carboxyanhydrides, for example N^{ε}-stearoyl-lysine-NCA, undergo spontaneous segregation of the enantiomers into enantiomorphous 2-D crystalline domains at the surface of water [194]. Polymerization reactions within such enantiomorphous crystallites, using nickel acetate as

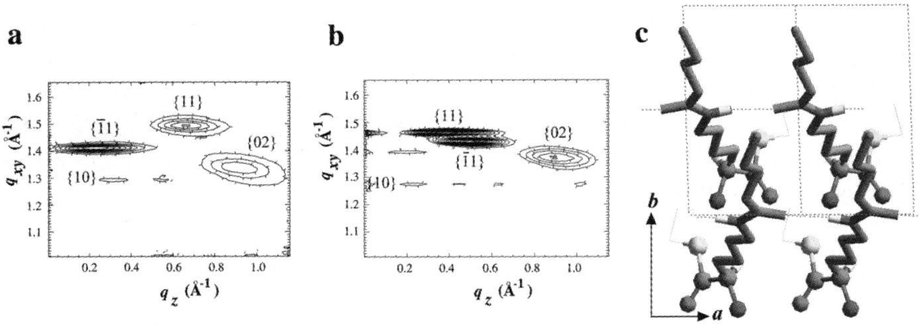

Fig. 15 **a, b** GIXD patterns measured from self-assembled 2-D crystallites of enantiomerically pure and racemic C_{18}-TE-Lys on water at 4 °C; **c** 2D packing arrangement of the racemic crystallites viewed perpendicular to the water surface. For clarity only part of the hydrocarbon chains is shown

catalyst, have yielded mixtures of oligopeptides up to six units long and of various enantiomeric compositions, as detected by MALDI-TOF MS using enantiolabeled monomers. Whereas the dipeptides and possibly the tripeptides display a random distribution, the tetra-, penta-, and hexapeptides exhibit an enhanced relative abundance of the homochiral sequences (by a factor of 2 to 3.5) compared to the binomial distribution [194].

The formation of racemic mixtures of homochiral oligopeptides is not confined to racemates undergoing spontaneous segregation into enantiomorphous domains; it can also be extended to racemic 2-D crystallites, provided the reaction pathway takes place preferentially between homochiral molecules related by translation symmetry, as in the case of the thioethylester of N^{ε}-stearoyl-lysine (C_{18}-TE-Lys) [193]. It has been experimentally proven by GIXD that racemic C_{18}-TE-Lys self-assemble into racemic 2-D crystallites, Fig. 15. Moreover, MALDI-TOF MS analyses of oligopeptide samples obtained from the polycondensation of deuterium enantiolabeled monomers have revealed a clear trend toward enhanced formation of di- to hexa-peptides with homochiral sequences, in agreement with a reaction pathway between homochiral monomer molecules related by translation symmetry [193, 195].

5.2 Enantiopure Oligopeptides from Non-Racemic Precursors

Racemic and enantiomorphous 2-D and 3-D crystals display different physical and chemical properties. This difference has been utilized to enhance chirality in non-racemic systems that self-assemble in racemic and enantiomorphous crystallites. Morowetz [196] has elaborated a mathematical model that considers an evaporation/crystallization process where the racemate is less soluble than the pure enantiomorphous crystal and the enantiomer (in excess) is concentrated in the solution. A similar enrichment of chirality has been achieved through sublimation experiments [197]. Another example of enantiomeric enrichment was reported for non-racemic esters of 9-anthroic acid by photoirradiation of mixtures of racemic and enantiomorphous crystals. The racemic crystals yielded an insoluble dianthracene dimer, while the enantiomorphous crystals are light stable and could be easily separated in high *ee* by solvent extraction, Scheme 9 [198].

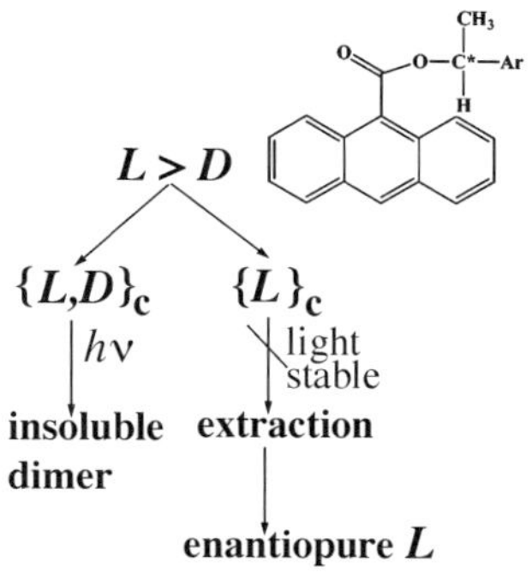

Scheme 9

Recently, the difference in structure and chemical reactivity between 2-D racemic and enantiomorphous crystallites has been used to generate enantiopure homochiral oligopeptides from non-racemic mixtures of amphiphilic α-amino acid NCAs. The racemic N^{α}-carboxyanhydride of γ-stearyl-glutamic acid (C_{18}-Glu-NCA) self-assembles on water to form racemic 2-D crystallites (Fig. 16a,b), as proved by GIXD.

According to the packing arrangement shown in Fig. 16c, it was anticipated that a lattice-controlled polymerization within such crystallites, in

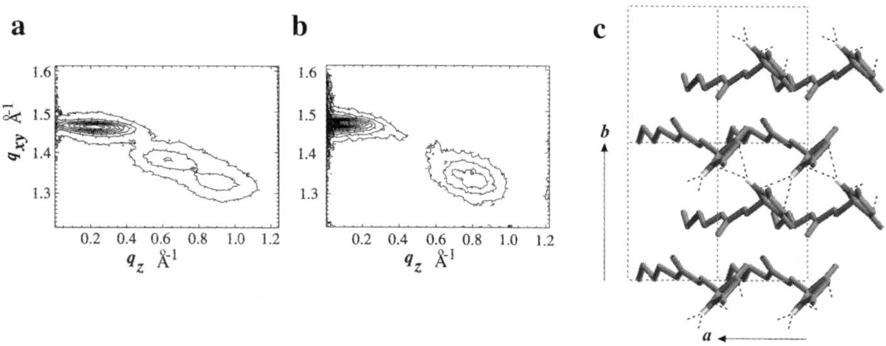

Fig. 16 a, b GIXD patterns of self-assembled 2-D crystallites of enantiomerically pure and racemic C_{18}-Glu-NCA on water; **c** The packing arrangement of the racemic 2-D crystallites viewed perpendicular to the water surface. For clarity, part of the hydrocarbon chains is not shown

contrast to those formed within racemic C_{18}-TE-Lys, would take place preferentially between glide-symmetry related heterochiral molecules via a zipper-like mechanism, to yield syndiotactic polymers, as indeed confirmed experimentally by MALDI-TOF MS, Fig. 17a [193].

When starting from chiral non-racemic mixtures of 3 : 7 and 4 : 6 L/D compositions, the short oligopeptides generated are rich in heterochiral diastereoisomers, whereas the longer oligomers are rich with oligopeptides of

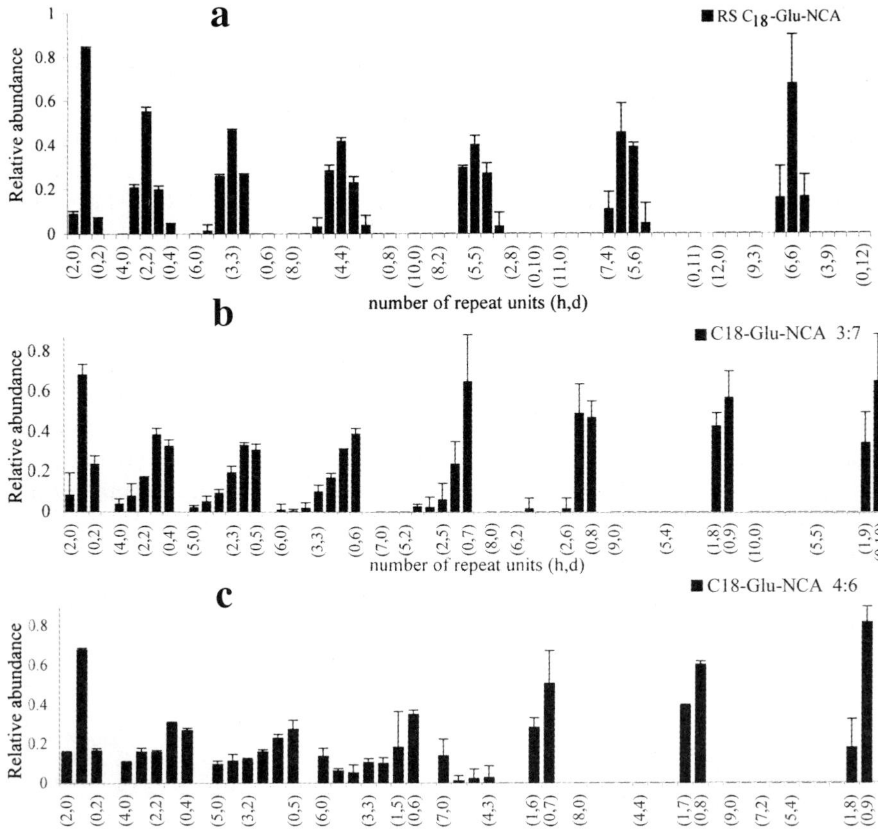

Fig. 17 MALDI-TOF MS analysis of oligopeptides obtained from **a** racemic and **b**, **c** 3:7 and 4:6 L/D non-racemic mixtures of C_{18}-Glu-NCA monomers. The vertical axis represents the relative abundance of each type of oligopeptide (h, d), where h is the number of R (unlabeled) repeat units and d the number of S (deuterated) repeat units; e.g. (4,0) is the tetrapeptide containing four D repeat units and zero L repeat units. For oligopeptides with the same number of repeat units, ion intensity (I) and amount are reliably proportional. The relative abundance was calculated according to the equation shown below for the (4,0) tetrapeptide: relative abundance $(4,0) = I(4,0)/\mathrm{I}[(4,0) + (3,1) + (2,2) + (1,3) + (0,4)]$. For clarity, the distributions of only some of oligopeptides are shown

single handedness. Figure 17b and c show that starting from a 3 : 7 ratio of L/D monomers, we obtained mixtures of diatereoisomeric oligopeptides nine and ten units long composed of only (1,8) and (0,9) L,D repeating units, respectively. Similarly, 4 : 6 L/D mixtures of monomers yielded oligopeptides eight and nine units long of only (1,7) and (0,8) L,D repeating units [199].

5.3
Homochiral Oligopeptides in a Phospholipid Environment

Vesicle and micelles are considered to be useful models for "minimum protocells" that had emerged in prebiotic times [200]. One of their properties should have been to sequester other molecules, including macromolecules, for self-replication. A central enigma to be addressed is related to various routes by which the enantiopure homochiral biopolymers were formed within such architectures. Polymerization of NCA of natural hydrophobic amino acids in water in the presence of phospholipids by Luisi et al. [201] has demonstrated that the hydrophobic environment enhances their rate of polymerization.

One possible pathway for the formation of homochiral biopolymers is achieved by embedding the crystalline architectures of non-racemic amphiphilic α-amino acids within a membrane-like environment. Using GIXD, it could be experimentally proven that non-racemic γ-stearyl-glutamic acid embedded within a DPPE phospholipid monolayer at the air/water interface underwent a phase separation into racemic and enantiomorphous crystallites [103]. Moreover, MALDI-TOF MS analyses of the oligopeptides generated from deuterium enantiolabeled non-racemic monomers have shown that polycondensation was initiated by the amine group of the DPPE molecules at the periphery of the monomer crystallites, yielding a preferential formation of oligopeptides of homochiral sequences, in keeping with a phase separation of the non-racemic monomers into the racemic and enantiomorphous 2-D crystallites, Scheme 10 [202].

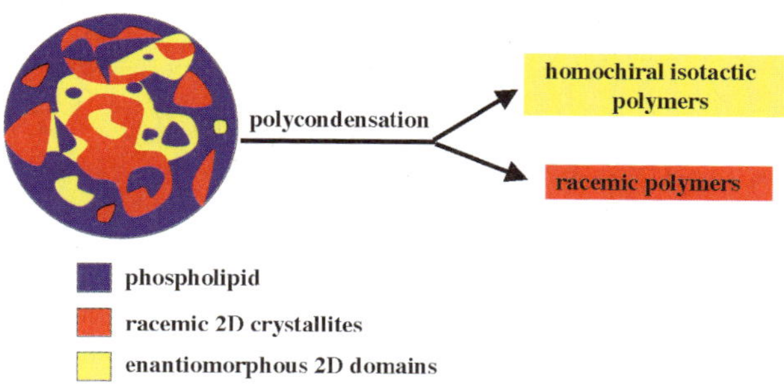

Scheme 10

5.4 Enantiopure Homochiral Oligopeptides Generated by Topochemical Reactions in 3-D Crystals

The solid-state polymerizations of several racemic optically-resolved amino acid NCAs were investigated by Kanazawa et al. [203, 204], who demonstrated, by kinetic and crystallographic investigations, that the rate of polymerization of these systems depends upon the packing arrangement of the monomers.

According to the packing arrangement of DL-NCA of the phenylalanine (PheNCA) crystal [205], Fig. 18, it was anticipated that a polymerization re-

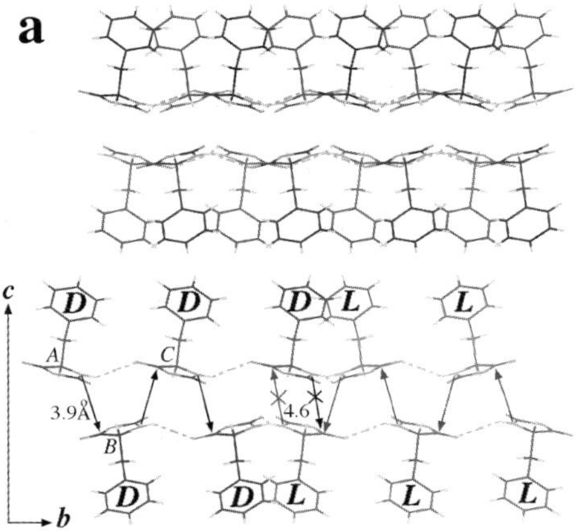

Fig. 18 Packing arrangement of (DL)-PheNCA crystals viewed along the *a*-axis. For clarity, some of the molecules are not shown. The reaction pathway for D and L molecules is shown with arrows

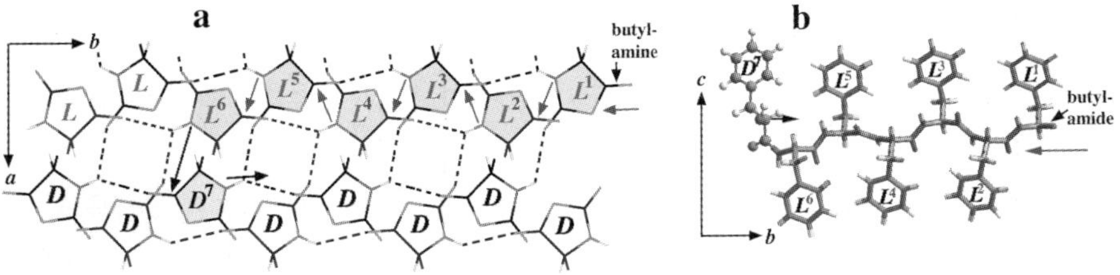

Fig. 19 **a** Packing arrangement of (DL)-PheNCA, viewed down the *c*-axis, showing the proposed chain termination by enantiomeric cross-inhibition of a (L_6) hexa-peptide, L^1 to L^6 (gray), with molecule D^7 (gray) to yield heptapeptides of sequence $Bu\text{-}(L)_{n-1} - D$. The unfavorable conformation after addition of D^7 unit would cause a rotation at the propagating end of the heptapeptide that would further enforce a flip in the direction of the chain propagation, thus implementing a chain termination. The modeled conformation of such a $Bu\text{-}(L)_6 - D$ heptapeptide is shown in **b**. The arrows indicate the direction of chain propagation

action should proceed preferentially between molecules of the same handedness [206], via a "zipper-like" mechanism coupled with an enantiomeric cross-inhibition, resulting in the formation of oligopeptides of homochiral sequences, Fig. 19.

Indeed, MALDI-TOF MS analyses of the oligopeptide products demonstrated the preferential formation of racemic mixtures of oligopeptides with homochiral sequences, Fig. 20, generated from deuterium enantiolabeled racemic monomer [206]. The degree of stereospecificity observed in this reaction increased as the homochiral oligopeptide length increased, as shown in Fig. 21.

X-ray powder diffraction and FTIR measurements of the reaction products indicated anti-parallel β-sheets formation, in agreement with our proposed polymerization mechanism, where the homochiral oligopeptide products should self-assemble into alternating poly-L- and poly-D-chains.

Reactivity within (DL)-PheNCA crystals provides a number of simple ways to de-symmetrize the racemic mixtures of the homochiral oligopeptides. For example, L-2-(thienyl)-alanineNCA (ThieNCA) molecules have been shown to enantioselectively occupy the L-sites in the DL-PheNCA host crystals. Lattice-controlled polymerization of such D-Phe/(L-Phe:*L*-Thie)-NCA mixed crystals yields libraries of non-racemic oligopeptides of ho-

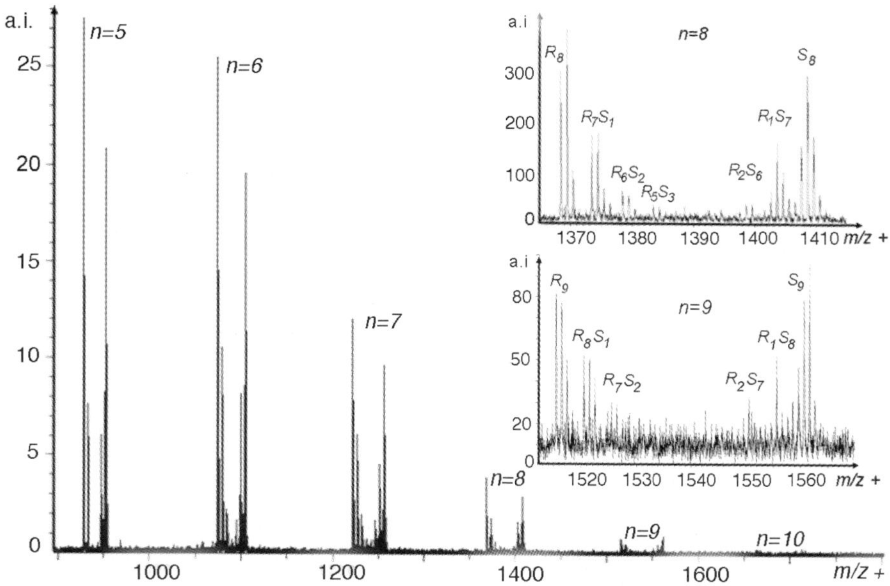

Fig. 20 MALDI-TOF mass spectrum of the oligopeptides obtained in the polymerization of (DL)-PheNCA at 22 °C, showing the m/z range from penta- to decapeptides. The two insets show expanded spectra of the octa- and nonapeptide ranges. The peaks at the wings of each group showing peptides of the same length represent molecules of homochiral sequence

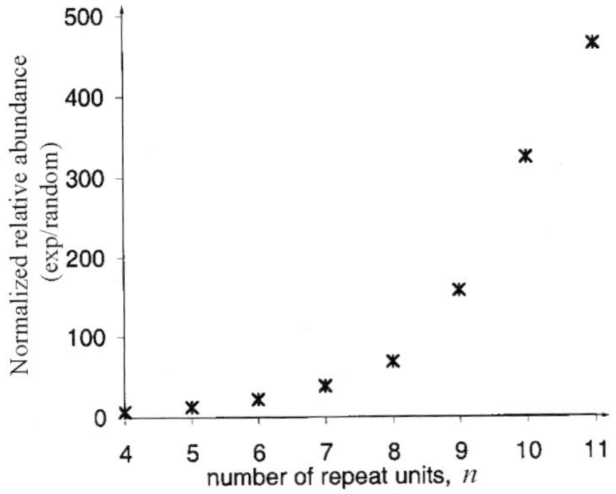

Fig. 21 Enhancement of the experimental relative abundance of the homochiral oligopeptides normalized to that calculated for a theoretical random process for molecules of any length n

mochiral sequences composed from mono component isotactic peptides of D-Phe repeating units and bi- or multi-component isotactic copolymers of (L-Phe : *L*-Thie) repeating units [207]. MALDI-TOF mass spectrometry of these three-component systems demonstrated that the L-Phe- and L-Thie-repeating units are randomly distributed within the copolymers. As a result of this random distribution, the departure from the non-racemic composition varies with chain length and the starting composition of the monomer mixture. In the overall product, all the oligopeptides containing one or more Thie- repeating units are enantiopure and the oligopeptides of homochiral Phe-sequences are enantiomerically enriched.

The above mechanism for enantioselective insertion of guest NCA amino acids within polymeric chains suggests a plausible scenario for the generation of libraries of diastereoisomeric mixtures of peptides starting from racemic mixtures of PheNCA as host and in the presence of racemic mixtures of other NCA amino acids. The L-guest molecules should occupy the L-sites in the host crystal, whereas the D-guests will occupy the D-sites. At regimes where the number of guest molecules is not sufficient to populate all possible sites in the chains of the oligopeptides, one can end up with a complex library of diastereoisomeric mixtures of peptides rather than with racemic ones. This mechanism of spontaneous symmetry breaking has some features in common with related mechanisms proposed recently [5, 208, 209]. Eschenmoser et al. [208] suggested, as part of his study on the self-assembly of higher oligomers of pyranosyl-RNA by ligative oligomerization of tetra-nucleotide-2′, 3′-cyclophosphates, that racemic mixtures containing all possible diastereisomeric sets can be expected to co-oligomerize stochastically and generate homochiral D- and L-oligomers predominantly. They also demonstrated that a true racemic mixture of the oligonucleotides is not possible after reaching a given length.

6 Self-Replication of Biopolymers

Several examples of non-enzymatic autocatalytic self-replicating systems based on template-directed synthesis of oligonucleotides have been reported [210–214] and recently reviewed [215]. Studies by Ghadiri et al. [216–218] have proven the feasibility of amplifying the chirality of oligopeptides by a self-replicating mechanism in solution through experiment. The group reported the design of an autocatalytic reaction between two short chiral peptides bearing sixteen repeating units of the same handedness, half of them electrophilic and half of them nucleophilic, that are properly assembled by non-covalent bonds on a longer peptide composed from 32 repeating units of the same handedness and related sequences. The product of the coupling of the two short peptides has exactly the same handedness and sequence as the longer peptide, and so this product can be used as a template for additional

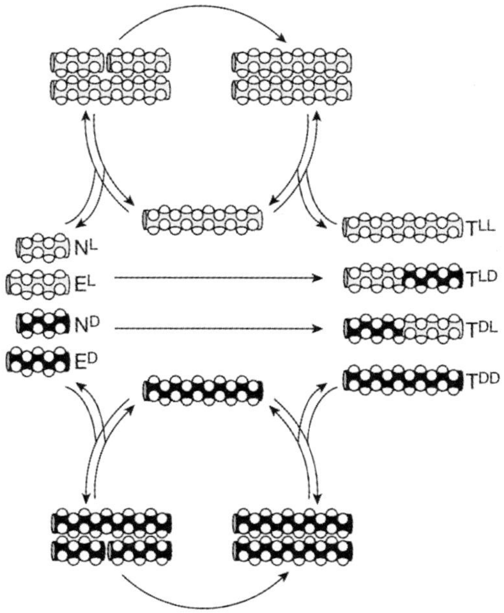

Fig. 22 A schematic representation of chiroselective replication cycles. Homochiral peptides T^{LL} and T^{DD} are produced autocatalytically while the heterochiral peptides T^{DL} and T^{DL} result from uncatalyzed background reactions. Template-directed ligation reactions proceed through the intermediary of stereospecific noncovalent complexes and pass on stereochemical information from the homochiral products to the substrates, thus resulting in the amplification of homochiral products. Light and dark backgrounds denote regions of the sequence composed of L- and D-amino acids, respectively (Reproduced from [218]. Copyright 2001, Nature)

self-replication. The results indicate that peptides, like nucleic acids, can operate as self-replicating systems and therefore perpetuate homochirality. The autocatalytic efficiency of this process is high, since a left-handed template is efficient at bringing together only those fragments that are also left-handed, Fig. 22.

The authors also showed that this system possesses very high fidelity: autocatalysis is significantly diminished if only one out of the 15 amino acid repeating units in the short peptide has opposite handedness. In this experiment, as in the organometallic systems described by Soai et al., the origin of the autocatalytic effect is augmented due to the fact that the long peptides composed from segments of opposite handedness, resulting from the reactions in solution, do not interfere with the autocatalytic process that takes place on the template.

7 Conclusions

In the absence of reliable fossils it is difficult or even impossible to address the historical question of how, and via which specific routes, homochirality emerged on Earth. Therefore, in this regard, we can only provide logical models that can outline scenarios of how the transition from racemic chemistry to homochiral biology might have happened.

Early theoretical models on the feasibility of stochastic mirror symmetry breaking at prebiotic conditions have been successfully realized under laboratory conditions, particularly in studies in crystal and surface science, asymmetric autocatalysis and polymer chemistry. The first step, common in all these scenarios, is the self-assembly of non-chiral or chiral molecules to form diastereoisomeric supramolecular architectures that display different physico-chemical properties.

Concentration of the organic reactants on surfaces or in the pores of clay materials prior to reaction has been suggested by Bernal [219] and Cairns-Smith [220]. Pores of different sizes might have operated as prebiotic reactors for asymmetric synthesis, since within their confined environment one may find chiral catalytic sites as well as chiral surfaces. One could envisage that such pores might have provided a plausible environment for the formation of diastereoisomeric self-assemblies of the types described in this review and as required for the stochastic mirror symmetry breaking scenarios. In addition, within such pores the chiral material once formed would be protected from racemization that could have been induced by impact with heavy bodies or by intense cosmic radiation.

Racemic mixtures do not exist in reality, since the number of molecules of the two enantiomers is never exactly equal. It has been calculated that a solution of a racemate has, on average, a fluctuation of 10^{12} molecules per mole [221]. Such fluctuations could direct reactions towards products of single handedness using autocatalytic Soai systems. In principle, such minute fluctuations should also be sufficient to amplify chirality via crystallization, but such processes have so far not been confirmed experimentally. In this respect, several laboratories have reported the presence of a bias in experiments involving "absolute" asymmetric synthesis in the solid phase; however, it has not been demonstrated that the origin of these effects is due to fluctuations from the racemic state or to artifacts resulting from chiral impurities present in the environment.

It has been suggested that weak interactions could be responsible for driving racemates towards homochirality via a deterministic process. However, it is difficult to deduce conclusions regarding the role played by these forces in chemical reactions for ensembles of molecules, since they induce a chiral bias of only 10^6 molecules per mole: six orders of magnitude lower than the stochastic fluctuations present in a racemate.

Efficient routes for the generation and amplification of homochiral peptides from racemic or from non-racemic monomers of low enantiomeric imbalance via polymerization in organized systems or on templates have been demonstrated. Homochiral oligopeptides have been shown to serve as auxiliaries for asymmetric synthesis and they can propagate and amplify their handedness since, once formed, such molecules can propagate their chirality to additional systems [222, 223]. Future studies will likely focus on spontaneous mirror symmetry breaking in sugars and nucleic acids. In this respect, the recent report by Joyce et al. [224] on the formation of chiral crystals of ribose derivatives from a complex soup of sugars is of importance.

Finally, a possible discovery of chiral materials and primitive life in the universe might throw additional light on this question of the origin of mirror symmetry breaking at prebiotic times.

Acknowledgements We thank Dr. J. G. Nery, R. Eliash and I. Rubinstein who participated in some of the experiments reported and the Israel Science Foundation, G. M. J. Schmidt Minerva Center for financial support. This work was supported by the European COST-D27 program "Prebiotic Chemistry and Early Evolution".

References

1. Fox S (1957) J Chem Ed 34:472
2. Wald G (1957) Ann NY Acad Sci 69:352
3. Bada JL, Miller SL (1987) Biosystems 20:21
4. Joyce GF, Visser GM, van Boeckel CA, van Boom JH, Orgel LE, van Westrenen J (1984) Nature 310:602
5. Avetisov V, Goldanski V (1996) Proc Natl Acad Sci USA 93:11435
6. Quack M (2002) Angew Chem Int Edit 41:4618
7. Kondepudi DK, Nelson GW (1985) Nature 314:438

8. Mason SF (1984) Nature 311:19
9. Mason SF (1989) Croat Chem Acta 62:165
10. Wang W, Min W, Zhu C, Fang Y (2003) Phys Chem Chem Phys 5:4000
11. Shinitzky M, Nudelman F, Barda Y, Haimovitz R, Chen E, Deamer DW (2002) Origins Life Evol B 32:285
12. Barron L (2004) Molecular light scattering and optical activity. Cambridge University Press, Cambridge
13. Rikken GLJ, Raupach E (2000) Nature 405:932
14. Franck P, Bonner WA, Zare RN (2000) In: Keinan E, Schecter I (eds) Chemistry for the 21st century. Wiley-VCH, Weinheim, p 175
15. Bailey J, Chrystospomou A, Hough JH, Glendhill TM, McCall A, Clark S, Menard F, Tamura M (1998) Science 281:672
16. Cronin JR, Pizzarello S (1997) Science 275:951
17. Pizzarello S, Zolensky M, Turk KA (2003) Geochim Cosmochim Acta 67:1589
18. Balavoine G, Moradpour A, Kagan HB (1974) J Am Chem Soc 96:5152
19. Flores JJ, Bonner WA, Massey GA (1977) J Am Chem Soc 99:3622
20. Pizzarello S, Weber AL (2004) Science 303:1151
21. Crisma M, Moretto A, Formaggio F, Kaptein B, Broxterman QB, Toniolo C (2004) Angew Chem Int Edit 43:6695
22. Kondepudi DK, Asakura K (2001) Acc Chem Res 34:946
23. Cintas P (2002) Angew Chem Int Edit 41:1139
24. Podlech J (1999) Angew Chem Int Edit 38:477
25. Feringa BL, van Delden RA (1999) Angew Chem Int Edit 38:3418
26. Podlech J (2001) Cell Mol Life Sci 58:44
27. Bonner WA (1999) Origins Life Evol B 29:615
28. Avalos M, Babiano R, Cintas P, Jimenez J, Palacios J (2000) Chem Commun 11:887
29. Avalos M, Babiano R, Cintas P, Jimenez J, Palacios J (2004) Origins Life Evol B 34:391
30. Nicollis G, Prigogine I (1977) Self-organization of nonequilibrium systems. Wiley, New-York
31. McBride JM, Carter RL (1991) Angew Chem Int Edit 30:293
32. Frank FC (1953) Biochem Biophys Acta 11:459
33. Calvin M (1969) Chemical evolution. Oxford University Press, Oxford
34. Havinga E (1954) Biochem Biophys Acta 13:171
35. Newman ACD, Powell HM (1952) J Chem Soc :3747
36. Imai H, Oaki Y (2004) Angew Chem Int Edit 43:1363
37. Oaki Y, Imai H (2004) J Am Chem Soc 126:9271
38. Oaki Y, Imai H (2005) Langmuir 21:863
39. Yu S-H, Cölfen H, Tauer K, Antonietti M (2002) Nano Lett 2:941
40. Yu S-H, Cölfen H, Tauer K, Antonietti M (2005) Nature Mat 4:51
41. Green BS, Lahav M, Rabinovich D (1979) Acc Chem Res 12:191
42. Farina M, Audisio G, Natta G (1967) J Am Chem Soc 89:5071
43. Penzien K, Schmidt GMJ (1969) Angew Chem Int Edit 8:608
44. Elgavi E, Green BS, Schmidt GMJ (1973) J Am Chem Soc 95:2058
45. Addadi L, Lahav M (1978) J Am Chem Soc 100:2838
46. Addadi L, Lahav M (1979) J Am Chem Soc 101:2152
47. Addadi L, Lahav M (1979) Pure Appl Chem 51:1269
48. Addadi L, Berkowitch-Yellin Z, Weissbuch I, van Mil J, Shimon LJW, Lahav M, Leiserowitz L (1985) Angew Chem Int Edit 24:466
49. Scheffer JR, Xia W (2005) Top Curr Chem 254:233 (and references cited therein)
50. Toda F (2005) Top Curr Chem 254:1 (and references cited therein)

51. Sakamoto M (2005) Top Curr Chem 254:207 (and references cited therein)
52. Hashizume D, Ohashi Y (2000) J Phys Org Chem 13:415 (and references cited therein)
53. Koshima H (2000) J Mol Struct 552:111 (and references cited therein)
54. Vestergren M, Eriksson J, Hakansson M (2003) Chem Eur J 9:4678
55. Hakansson M (2004) J Organomet Chem 689:1723
56. Sato I, Kadowaki K, Urabe H, Jung JH, Ono Y, Shinkai S, Soai K (2003) Tetrahedron Lett 44:721
57. Addadi L, Berkovitch-Yellin Z, Domb N, Gati E, Lahav M, Leiserowitz L (1982) Nature 296:21
58. Weisinger-Lewin Y, Frolow F, McMullan RK, Koetzle TF, Lahav M, Leiserowitz L (1989) J Am Chem Soc 111:1035
59. Vaida M, Shimon LJW, Weisinger-Lewin Y, Frolow F, Lahav M, Leiserowitz L, McMullan R (1988) Science 241:1475
60. Vaida M, Shimon LJW, van Mil J, Ernst-Cabrera K, Addadi L, Leiserowitz L, Lahav M (1989) J Am Chem Soc 111:1029
61. Weissbuch I, Addadi L, Lahav M, Leiserowitz L (1991) Science 253:637
62. Chenchaiah PC, Holland HL, Munoz B, Richardson MF (1986) J Chem Soc Perkin T 2:1775
63. Weissbuch I, Lahav M, Leiserowitz L, Meredith GR, Vanherzeele H (1989) Chem Mater 1:14
64. Gervais C, Wust T, Behrnd NR, Wubbenhorst M, Hulliger J (2005) Chem Mat 17:85
65. Kahr B, Gurney RW (2001) Chem Rev 101:893
66. McBride JM, Bertman SB (1989) Angew Chem Int Edit 28:330
67. McBride JM (1989) Angew Chem Int Edit 28:377
68. Welch C (2001) Chirality 13:425
69. Weissbuch I, Popovitz-Biro R, Lahav M, Leiserowitz L (2000) In: Mersmann A (ed) Handbook of crystallization. Marcel Dekker, New York, p 401
70. Weissbuch I, Berkovitch-Yellin Z, Leiserowitz L, Lahav M (1985) Israel J Chem 25:362
71. Hazen RM, Filley TR, Goodfriend GA (2001) Proc Natl Acad Sci USA 98:5487
72. Hazen RM, Sholl DS (2003) Nature Mat 2:367
73. Cody AM, Cody RD (1991) J Cryst Growth 113:508
74. Orme CA, Noy A, Wierzbicki A, McBride MT, Grantham M, Teng HH, Dove PM, DeYoreo JJ (2001) Nature 411:775
75. McFadden CF, Cremer PS, Gellman AJ (1996) Langmuir 12:2483
76. Attard GA, Ahmadi A, Feliu JM, Rodes A, Herrero E, Blais S, Jerkiewcz J (1999) J Phys Chem B 103:1381
77. Horvath JD, Gellman AJ, Sholl DS, Power TD (2002) In: Hicks JM (ed) Chirality: Physical chemistry (ACS Symp Ser 810), Ch 19. Oxford University Press, Oxford, p 269
78. Kuzmenko I, Weissbuch I, Gurovitz I, Leiserowitz L, Lahav M (1998) Chirality 10:415
79. Lahav M, Leiserowitz L (1999) Angew Chem Int Edit 38:2533
80. Feyter SD, Gesqiere A, Wurst K, Amabilino DB, Veciana J, Schryver FCD (2001) Angew Chem Int Edit 40:3217
81. Walba DM, Stevens F, Clarck NA, Parks DC (1996) Acc Chem Res 29:591
82. Ernst KH, Kuster Y, Fasel R, Muller M, Ellerbeck U (2001) Chirality 13:675
83. Böhringer M, Morgenstern W-D, Schneider R, Berndt R (1999) Angew Chem Int Edit 38:821
84. Bohringer M, Morgenstern K, Schnider W-D, Berndt R, Mauri F, DeVita A, Car R (1999) Phys Rev Lett 83:324

85. Chen Q, Frenkel DJ, Richardson NV (2002) Langmuir 18:3219
86. Smith DPE (1991) Vacuum Sci Technol B 9:1119
87. Sowerby SJ, Heckel WM, Petersen GB (1996) J Mol Evol 43:419
88. Kuhnle A, Linderoth TR, Hammer B, Besenbacher F (2002) Nature 415:891
89. Chen Q, Richardson NV (2003) Nature Mat 2:324
90. Moffatt DJ, Lopinski GP, Wayner DDM, Wolkow RA (2002) In: Hicks JM (ed) Chirality: Physical chemistry (ACS Symp Ser 810). Oxford University Press, Oxford, p 283
91. Lundquist M (1978) Prog Chem Fats Other Lipids 16:101
92. Stewart MV, Arnett EM (1982) In: Elliel EL, Wilen SH, Allinger NL (eds) Topics in stereochemistry, vol 13. Wiley, New York, p 195
93. Weiss RM, McConnell HM (1984) Nature 310:47
94. Rietz R, Brezesinski G, Mohwald H (1993) Ber Bunsenges Phys Chem 97:1394
95. Nandi N, Volhardt D (2003) Chem Rev 103:4033
96. Gruniger H, Möbius D, Meyer H (1983) J Chem Phys 79:3701
97. Orrit M, Möbius D (1986) J Chem Phys 85:4966
98. Loschek R, Möbius D (1988) Chem Phys Lett 151:176
99. Hönig D, Möbius D (1991) J Phys Chem 95:4590
100. Hoffmann F, Stine KJ, Huhnerfuss H (2005) J Phys Chem B 109:240
101. Kuzmenko I, Rapaport H, Kjaer K, Als-Nielsen J, Weissbuch I, Lahav M, Leiserowitz L (2001) Chem Rev 101:1659
102. Weissbuch I, Berfeld M, Bouwman WG, Kjaer K, Als-Nielsen J, Lahav M, Leiserowitz L (1997) J Am Chem Soc 119:933
103. Weissbuch I, Rubinstein I, Weygand MJ, Kjaer K, Leiserowitz L, Lahav M (2003) Helv Chim Acta 86:3867
104. Nassoy P, Goldmann M, Bouloussa O, Rondelez F (1995) Phys Rev Lett 75:457
105. Leveiller F, Jacquemain D, Lahav M, Leiserowitz L, Deutsch M, Kjaer K, Als-Nielsen J (1991) Science 252:1532
106. Leveiller F, Böhm C, Jacquemain D, Möhwald H, Leiserowitz L, Kjaer K, Als-Nielsen J (1994) Langmuir 10:819
107. Viswanathan R, Zadadzinski JA, Schwartz DK (1994) Nature 368:440
108. Jiao T, Liu M (2005) J Phys Chem B 109:2532
109. Yuan J, Liu M (2003) J Am Chem Soc 125:5051
110. Zhang L, Lu Q, Liu M (2003) J Phys Chem B 107:2565
111. Huang X, Li C, Jiang S, Wang X, Zhang B, Liu M (2004) J Am Chem Soc 126:1322
112. Huang X, Liu M (2003) Chem Commun :66
113. Huang X, Jiang S, Liu M (2004) J Phys Chem B 109:114
114. Cornelissen JJLM, Rowan AE, Nolte RJM, Sommerdijk NAJM (2001) Chem Rev 101:4039
115. Mateos-Timoneda MA, Crego-Calama M, Reinhoudt DN (2005) Supramol Chem 17:67
116. Hanan GS, Lehn J-M, Krytsakas N, Fisher J (1995) Chem Commun :765
117. Bassani DM, Lehn J-M, Baum G, Fenske D (1997) Angew Chem Int Edit 36:1845
118. van Esch J, De Feyter S, Kellogg RM, De Schrijver F, Feringa BL (1997) Chem Eur J 3:1238
119. Kirstein S, von Berlepsch H, Bottecher C, Ouart A, Reck G, Dahne S (2003) ChemPhysChem 3:146
120. Ribo JM, Crusats J, Sagues F, Claret J, Rubires R (2001) Science 292:2063
121. Rubires R, Farrera J-A, Ribo JM (2001) Chem Eur J 7:436

122. Crusats J, Claret J, Diez-Perez I, El-Hachemi Z, Garcia-Ortega H, Rubires R, Sagues F, Ribo JM (2003) J Chem Soc Chem Commun 1588
123. Yamaguchi T, Kimura T, Matsuda H, Aida T (2004) Angew Chem Int Edit 43:6350
124. Lauceri R, Raudino A, Scolaro LM, Mical N, Purrello R (2002) J Am Chem Soc 124:894
125. Takats Z, Nanita SC, Cooks RG (2003) Angew Chem Int Edit 42:3521
126. Julian R, Hodyss R, Kinnear B, Jarrold M, Beauchamp JL (2002) J Phys Chem B 106:1219
127. Julian RR, Myung S, Clemmer DE (2004) J Phys Chem B 108:6105
128. Myung S, Julian RR, Nanita SC, Cooks RG, Clemmer DE (2004) J Am Chem Soc 126:4110
129. Green BS, Heller L (1974) Science 185:525
130. van Mil J, Gati E, Addadi L, Lahav M (1981) J Am Chem Soc 103:1248
131. van Mil J, Addadi L, Lahav M, Boyle WJ, Sifniades S (1987) Tetrahedron 43:1281
132. Weissbuch I, Addadi L, Berkovitch-Yellin Z, Gati E, Lahav M, Leiserowitz L (1984) Nature 310:161
133. Weissbuch I, Frolow F, Addadi L, Lahav M, Leiserowitz L (1990) J Am Chem Soc 112:7718
134. Weissbuch I, Popovitz-Biro R, Leiserowitz L, Lahav M (1994) In: Behr J-P (ed) Perspectives in supramolecular chemistry, vol 1: The state of the art—100 years of the lock-and-key principle. Wiley, New York, p 173
135. Weissbuch I, Leiserowitz L, Lahav M (2002) In: Hicks JM (ed) Chirality: Physical Chemistry (ACS Symp Ser 810), Ch 17. Oxford University Press, Oxford, p 242
136. Torbeev V, Weizmann Institute of Science, unpublished results
137. Weissbuch I, Leiserowitz L, Lahav M (2003) Cryst Growth Des 3:125
138. Kondepudi DK, Kaufman RJ, Singh N (1990) Science 250:975
139. McBride JM, Carter RL (1991) Angew Chem Int Edit 30:293
140. Metcalf G, Ottino JM (1994) Phys Rev Lett 72:2875
141. Martin B, Tharrington A, Wu X-l (1996) Phys Rev Lett 77:2826
142. Viedma C (2004) J Cryst Growth 261:118
143. Viedma C (2005) Phys Rev Lett 94:065504
144. Wynberg H, Feringa BL (1976) Tetrahedron 32:2831
145. Girard C, Kagan H (1998) Angew Chem Int Edit 37:2922
146. Blackmond DC (2000) Acc Chem Res 33:402
147. Blackmond DG, McMillan CR, Ramdeehul S, Schorm A, Brown JM (2001) J Am Chem Soc 123:10103
148. Blackmond DG (2004) Proc Natl Acad Sci USA 101:5732
149. Mathew SP, Iwamura H, Blackmond DG (2004) Angew Chem Int Edit 43:2099
150. Avalos M, Babiano R, Cintas P, Jimenez JL, Palacios JC (1997) Tetrahedron Asymmetr 8:2997
151. Soai K, Niwa S, Hori H (1990) J Chem Soc Chem Comm :982
152. Soai K, Shibata T, Sato I (2000) Acc Chem Res 33:382
153. Sato I, Urabe H, Ishiguro S, Shibata T, Soai K (2003) Angew Chem Int Edit 42:315
154. Soai K, Shibata T, Kowata Y (1997) Japan Kokai Tokkyo Koho 9,268,179 (application date: February 1, 1996 and April 18, 1996)
155. Soai K, Sato I, Shibata T, Komiya S, Hayashi M, Matsueda Y, Imamura H, Hayase T, Morioka H, Tabira H, Yamamoto J, Kowata Y (2003) Terahydron Asymmetry 14:185
156. Kawasaki T, Sato M, Ishiguro S, Saito T, Morishita Y, Sato I, Nishino H, Inoue Y, Soai K (2005) J Am Chem Soc 127:3274
157. Singleton DA, Vo LK (2003) Org Lett 5:4337

158. Soai K, Osanai S, Kadowaki K, Yonekubo S, Shibata T (1999) J Am Chem Soc 121:11235
159. Sato I, Kadowaki K, Ohgo Y, Soai K (2004) J Mol Catal Chem 216:209
160. Sato I, Sugie R, Matsueda Y, Furumura Y, Soai K (2004) Angew Chem Int Edit 43:4490
161. Lahav N, White D, Chang S (1978) Science 210:67
162. Plankensteiner K, Righi A, Rode BM (2002) Origins Life Evol B 32:225
163. de Duve C (2002) Life evolving: Molecules mind and meaning. Oxford University Press, New York
164. Wächtershäuser G (1988) Microbiol Rev 52:452
165. Huber C, Wächtershäuser G (1997) Science 276:245
166. Huber C, Eisenreich W, Hecht S, Wächtershäuser G (2003) Science 301:938
167. Leman L, Orgel L, Ghadiri MR (2004) Science 306:283
168. Commeyras A, Collet H, Boiteau L, Taillades J, Trambouze O, Cottet H, Biron J-P, Plasson R, Mion L, Lagrille O, Martin H, Selsis F, Dobrijevic M (2002) Polymer Int 51:661
169. Vayaboury W, Giani O, Collet H, Commeyras A, Schue F (2004) Amino Acids 27:161
170. Blair NE, Bonner WA (1980) Origins Life Evol B 10:255 (and references cited therein)
171. Inoue S, Matsuura K, Tsuruta T (1968) J Polym Sci C23:271
172. Brack A, Spach G (1979) J Mol Evol 13:35
173. Goldberg SI, Crosby JM, Iusem ND, Younes UE (1987) J Am Chem Soc 109:823
174. Blocher M, Hitz T, Luisi PL (2001) Helv Chim Acta 84:842
175. Hitz T, Luisi PL (2002) Helv Chim Acta 85:3975
176. Hitz T, Luisi PL (2003) Helv Chim Acta 86:1423
177. Blair NE, Dirbas FM, Bonner WA (1981) Tetrahedron 37:27
178. Spach G, Brack A (1980) J Mol Evol 15:231
179. Bonner WA, Blair NE, Dirbas FM (1981) Origins Life 11:119
180. Green MM, Peterson NC, Sato T, Teramoto A, Cook R, Lifson S (1995) Science 268:1860
181. Wittung P, Nielsen PE, Buchardt O, Egholm M, Norden B (1994) Nature 368:561
182. Rowan AE, Nolte RJM (1998) Angew Chem Int Edit 37:63
183. Huck NPM, Jager WF, deLange B, Feringa BL (1996) Science 273:1686
184. Addadi L, van Mil J, Lahav M (1981) J Am Chem Soc 103:1249
185. van Mil J, Addadi L, Gati E, Lahav M (1982) J Am Chem Soc 104:3429
186. Addadi L, Weinstein S, Gati E, Weissbuch I, Lahav M (1982) J Am Chem Soc 104:4610
187. Zbaida D, Weissbuch I, Shavit-Gati E, Addadi L, Leiserowitz L, Lahav M (1987) Reactive Polymers 6:241
188. Zbaida D, Lahav M, Drauz K, Knaup G, Kottenhahn M (2000) Tetrahedron 56:6645
189. Fukuda K, Shibasaki Y, Nakahara H (1981) J Macromol Sci Chem A15:999
190. Fukuda K, Shibasaki Y, Nakahara H, Liu M (2000) Adv Colloid Interf Sci 87:113
191. Eliash R, Weissbuch I, Weygand MJ, Kjaer K, Leiserowitz L, Lahav M (2004) J Phys Chem B 108:7228
192. Shibata A, Hashimura Y, Yamashita S, Ueno S, Yamashita T (1991) Langmuir 7:2261
193. Zepik H, Shavit E, Tang M, Jensen TR, Kjaer K, Bolbach G, Leiserowitz L, Weissbuch I, Lahav M (2002) Science 295:1266
194. Weissbuch I, Bolbach G, Zepik H, Shavit E, Tang M, Frey J, Jensen TR, Kjaer K, Leiserowitz L, Lahav M (2002) J Am Chem Soc 124:9093
195. Weissbuch I, Zepik H, Bolbach G, Shavit E, Tang M, Jensen TR, Kjaer K, Leiserowitz L, Lahav M (2003) Chem Eur J 9:1782
196. Morowetz HJ (1969) J Theor Biol 25:491

197. Elliel E, Wilen S (1994) Stereochemistry of Organic Compounds. Wiley, New York
198. Lahav M, Laub F, Gati E, Leiserowitz L, Ludmer Z (1976) J Am Chem Soc 98:1620
199. Weissbuch I, Bolbach G, Leiserowitz L, Lahav M (2004) Origins Life Evol B 34:79
200. Deamer DW (1997) Microb Mol Biol Rev 61:239
201. Blocher M, Liu D, Walde P, Luisi PL (1999) Macromolecules 32:7332
202. Rubinstein I, Bolbach G, Weygand MJ, Kjaer K, Weissbuch I, Lahav M (2003) Helv Chim Acta 86:3851
203. Kanazawa H (1992) Polymer 33:2557
204. Kanazawa H, Ohashi Y (1996) Mol Cryst Liq Cryst 277:45
205. Kanazawa H, Uekusa H, Ohashi Y (1997) Acta Crystallogr C 53:1154
206. Nery JG, Bolbach G, Weissbuch I, Lahav M (2003) Angew Chem Int Edit 42:2157
207. Nery JG, Bolbach G, Weissbuch I, Lahav M (2005) Chem Eur J 11:3039
208. Bolli M, Micura R, Eschenmoser A (1997) Chem Biol 4:309
209. Siegel JS (1998) Chirality 10:24
210. von Kiedrowski G (1986) Angew Chem Int Edit 25:932
211. von Kiedrowski G (1993) Bioorg Chem Front 3:113
212. Orgel LE (1992) Nature 358:203
213. Severin D, von Kiedrowski G (1994) Nature :221
214. Li KC, Nicolaou T (1994) Nature 369:218
215. Paul N, Joyce GF (2004) Curr Opin Chem Biol 8:634
216. Lee DH, Granja JR, Martinez JA, Severin K, Ghadiri MR (1996) Nature 382:525
217. Lee DH, Severin K, Yokobayashi Y, Ghadiri MR (1997) Nature 390:591
218. Sagathelian A, Yokobyashi Y, Soltani K, Ghadiri MR (2001) Nature 409:797
219. Bernal JD (1969) The origin of life. World Publishing Co, Cleveland, OH
220. Cairns-Smith G (1986) Clay minerals and the origin of life. Cambridge University Press, Cambridge
221. Dunitz JD (1996) Proc Natl Acad Sci USA 93:14 260
222. Banfi S, Colona S, Molinari H, Julia S, Guixer J (1984) Tetrahedron 40:5207
223. Kelly DR, Meck A, Roberts SM (2004) Chem Commun 18:2018
224. Springsteen G, Joyce GF (2004) J Am Chem Soc 126:9578

Top Curr Chem (2005) 259: 167–211
DOI 10.1007/tcc001

Published online: 23 August 2005

Evolutionary Potential and Requirements for Minimal Protocells

Eörs Szathmáry[1,2] (✉) · Mauro Santos[1,3] · Chrisantha Fernando[1,4]

[1]Collegium Budapest, Institute for Advanced Study, Szentháromság u. 2, 1014 Budapest, Hungary
Szathmary@colbud.hu

[2]Research Group of Theoretical Biology and Ecology, Eötvös University, Pázmány Péter sétány 1/c, 1117 Budapest, Hungary
Szathmary@colbud.hu

[3]Departament de Genètica i de Microbiologia, Grup de Biologia Evolutiva, Universitat Autònoma de Barcelona, Bellaterra, 08193 Barcelona, Spain

[4]Department of Informatics, Center for Computational Neuroscience and Robotics, University of Sussex, Falmer BN1 9RH, UK

1	**Introduction**	168
1.1	Units of Evolution and Units of Life	168
1.2	Criteria for Minimal Life	170
2	**Compartmentation: Membranes, Reproducing Microspheres and the Lipid World**	173
2.1	Autocatalytic Membrane Formation	173
2.2	Membrane Growth and Inheritance	174
2.3	Vesicle Division	177
2.4	Compartment/Template Infrabiological Systems	179
2.5	Membrane Permeability	180
3	**Primordial Genomes**	183
3.1	Error Thresholds	183
3.2	Vesicle Models	184
3.3	Recombination (Sex) and Gene Redundancy	187
3.4	Lessons from "Bags of Genes" in Contemporary Genetic Systems	188
3.5	Minimal Set of Genes for Cellular Life: The Top-Down Approach	192
4	**Metabolism**	193
4.1	All Living Systems Today are Metabolic	193
4.2	Evolution of Autocatalytic Cycles	196
4.3	The Formose Cycle and the Reductive Citric Acid Cycle	199
4.4	Encapsulation of Autocatalytic Cycles	200
4.5	Post-Enzymatic Evolution of Metabolism	202
5	**Outlook**	207
References		208

Abstract Minimal protocell concepts have high intellectual and practical value. Following chemoton theory, developed by Tibor Gánti in 1971, we suggest that a minimal protocell satisfying all life criteria should consist of three subsystems: a metabolic network producing materials for the production of all three subsystems at the expense of the difference between food and waste products, a genetic subsystem based on template polymerization and a boundary subsystem consisting of a bilayer vesicle. All three subsystems are autocatalytic and the system as a whole (called chemoton) is also autocatalytic. The chemoton can undergo spatial reproduction in the right parameter domain. Various infrabiological systems can be constructed from any two subsystems; we review the experimental attempts aimed at such a goal. As the complexity of the realized systems increases, the problem of unwanted side reactions becomes more and more dramatic in non-enzymatic systems. No satisfactory solution to the problem of metabolite channeling is known.

Keywords Autocatalysis · Chemoton · Evolution · Heredity · Infrabiological systems · Lipid world · Metabolism · Minimal life · Protocell · Unit of evolution

Abbreviations

AL	artificial life
EC	Enzyme Commission
GARD	graded autocatalysis replication domain
PNA	peptide nucleic acid
SCM	stochastic corrector model

1 Introduction

1.1 Units of Evolution and Units of Life

Evolution by natural selection is perhaps the most important process acting in populations of living systems. This is one of the reasons why it is so tempting to equate units of evolution (i.e. an abstract generalisation that makes no reference whatsoever to any particular level of biological organisation) to units of life. Another reason is that units of evolution can be much more readily defined. There are a few known alternative formulations of the concept of units of evolution; here we stick to the version outlined by Maynard Smith [1, 2]: such units must multiply, show heredity across generations (like begets like), and heredity should not be exact. If some of the hereditary traits affect the chance of reproduction and/or survival of the units, evolution by natural selection can take place in a population of such units. The combination of survival and reproduction (translating into the expected number of descendants) is called fitness.

The above characterisation of Darwinian dynamics is deliberately general: note that it is not restricted to cover living systems only. (As a matter of fact some living systems do not—sometimes cannot—multiply: mules and neurons normally do not reproduce; we shall come back to this point.) Hence it is potentially applicable to molecules and cultural traits *as far as the criteria really apply.*

A general point about definitions is that they cannot be falsified. They have to be internally consistent, of course, but there can be an arbitrary number of such definitions for life, for example. It is the use of the alternative definitions that makes the difference: some definitions are found helpful because they categorise natural phenomena in a way that is conducive to further insights. There is always an ingredient of arbitrariness in definitions: we have to live with this fact.

Coming back to the problems of units of life, it is obvious that if we think that they are equivalent to units of evolution, then replicating RNA molecules or some computer viruses would be alive. The first option would be endorsed by some protagonists of the RNA world, whereas the second alternative would be maintained by some AL researchers. We do not find this view useful: in short, we do not think that all replicators *sensu* Dawkins [3] are living systems. We appreciate the desire on behalf of researchers to move the goalpost so that a "living molecule" would provide us with the solution to the problem of the origin of life, but this would not explain much about the origin of cells (or "cellular life", as some use to distinguish it from merely "molecular" life). Today every autonomous living system is cellular (prokaryotic or eukaryotic, uni- or multicellular) in nature, as advocated by the Schwann–Schleiden cell theory in the mid-nineteenth century. Viruses are not (autonomously) alive. Gánti's [4, 5] analogy is very useful: a virus is to a cell what a replicating programme is to a computer. The cell is happily alive without the virus; the virus cannot do anything on its own (apart from slow disintegration). Thus if we search for the principles of life *sensu* Gánti we have to search for principles describing cellular life. The sensible relation, by our definition, between units of evolution and units of life is thus that of two partially overlapping sets [6].

As Gánti emphasises, modelling of a living system in entire generality is not very rewarding, since there are (at least) two levels of life: multicellular organisms consist of units that are living systems even if the multicellular organism is killed. Conversely, the death of many of its cells does not necessarily kill the organism. If we are interested in the origin of life, our ultimate target must be the explanation of the origin of the prokaryotic (bacterial) cell.

But a prokaryotic cell is far too complex for spontaneous self-assembly or self-organisation. Even the simplest known living creatures (such as mycoplasms) contain several hundred genes: they must be products of past evolution. And we have to assume that sometime, more than 3 billion years ago, there were no genes at all. The increase in complexity must have come about by duplication and divergence, and merging (recombination) [7]. We conclude that a lot of evolution must have preceded the origin of bacterial cells. Some of this evolution may have taken place even in a pre-cellular phase of evolution (such as the naked RNA world [8]). And the simplest cells must have been very remote from contemporary instantiations; hence a lot of evolution must have happened between the protocellular and bacterial phases of evolutionary history.

1.2 Criteria for Minimal Life

With the caveats above, we feel ready to head towards a working definition of minimal life. We base this working definition on the chemoton concept [4]: a minimal living system is a chemical supersystem comprising three systems: a metabolic network, template replication and a boundary system. It is instructive to look at the abstract minimal version of the chemoton (Fig. 1). It is important that all three systems are autocatalytic, and via the shown stoichiometric coupling, the chemoton as a whole is also autocatalytic. As we shall discuss later, such a system is not only chemically autocatalytic but also biologically reproducing by cell division. This biological minimal system will be our intellectual starting point. It is true that nobody has ever claimed that the system in Fig. 1 can be realised by a handful of reactions. This merely reflects that the chemoton is the result of abstraction and idealisation. A pertinent question for the practical chemist is this: what is the requisite chemical complexity of a realistic chemoton? We do not know, but this question defines a research programme.

As mentioned above, this system is by definition not a general model of living systems: it applies (discounting a lot of evolutionary detail) to the bacterial level of organisation. But we should mention that it satisfies the so-called criteria for life, set up by Gánti. Life criteria are phenomenological and verbal, but rigorous descriptions of those traits that apply to all kinds of living organisms. It is important for our discussion that Gánti distinguishes absolute from potential life criteria: the former are necessary for all units of life, the latter are necessary only if *populations* of living systems are to survive and evolve. A mule is a unit of life, but not a unit of evolution.

It follows from our foregoing discussion that such a system must be a culmination of a protracted period of prior evolution. This comprises chemical evolution (the complexification of chemical systems) and evolution by natural selection of chemical replicators of various kinds. It is likely that mineral surfaces have played an important role in precellular evolution (e.g. [9–12]). Surfaces have favourable thermodynamic, kinetic and selective effects on chemical and replicator evolution. Reviews of molecular selection dynamics on surfaces can be found elsewhere [13]. We mention this link because effects that surfaces can confer can be conferred even more efficiently by compartments: obviously, a reproducing protocell is the strongest form of population structure, conducive to group selection [14, 15] of the replicators included within.

Although we take the chemoton as the working definition of minimal life, it is important that its three subsystems can be combined to yield three different

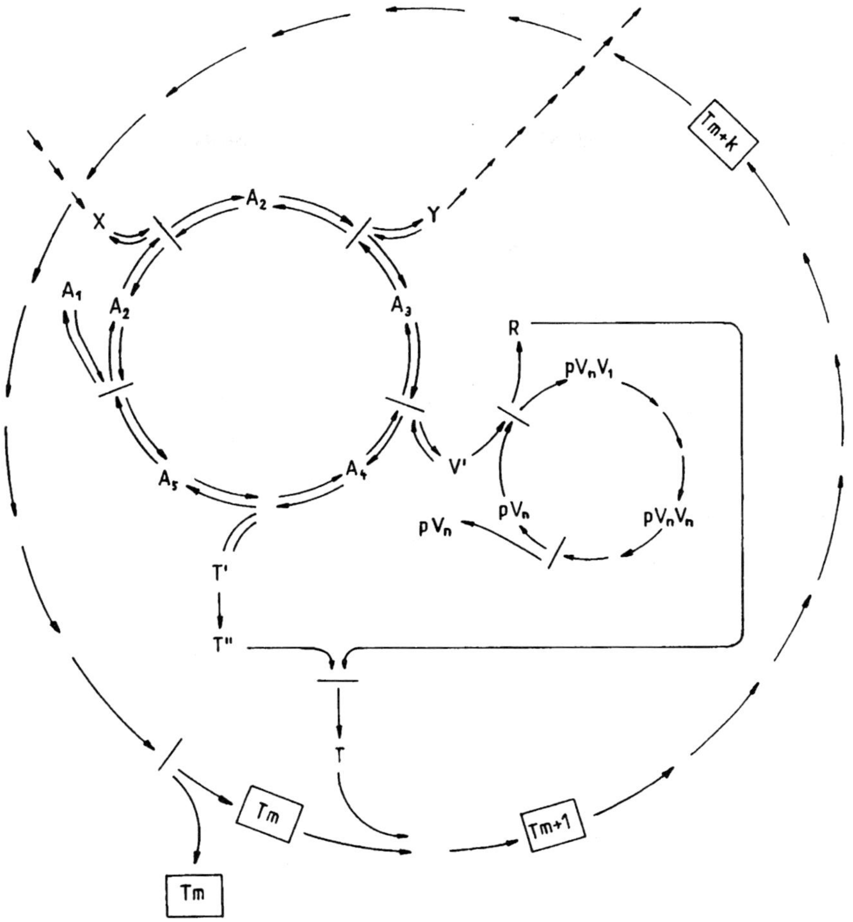

Fig. 1 Gánti's abstract chemoton model for minimal life. A_i are intermediates of the metabolic cycle, pV_n is a template molecule consisting of n pieces of monomer V, V′ is the activated monomer, T′ is the precursor to the membranogenic molecule T. T_m denotes a membrane consisting of m pieces of T. The system works at the expense of the difference between X and Y. Note that all three subsystems are autocatalytic

doublet systems (Fig. 2). It is historically interesting to note that the original formulation of a chemoton [4] consisted of the metabolic cycle and the replicating template only. By definition such a system is not actively compartmentalised, because it is lacking a self-generated boundary system, although it could have played a major role in precellular evolution (e.g. [13]). The combination of a metabolic cycle and a membrane was conceived also by Gánti [16], and called a self-reproducing microsphere. In contrast, Szostak et al. [17] conceived a protocell-like entity with a boundary and template replication but no metabolic subsystem.

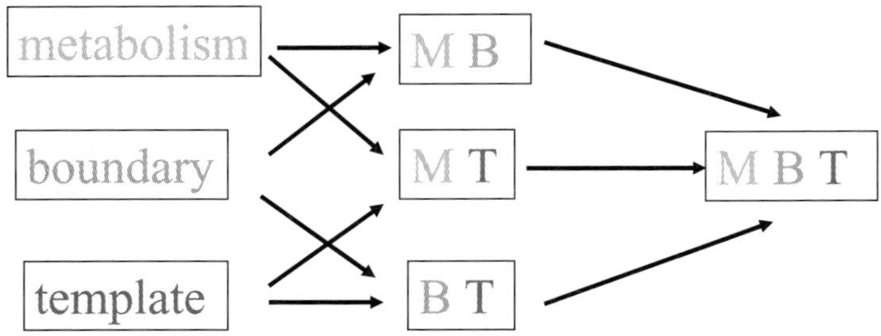

Fig. 2 Elementary combinatorics of infrabiological systems. The chemoton is a biological minimal system comprising three qualitatively different subsystems (metabolism, membrane, and template)

Similar criteria for a protocell were suggested by Pohorille and Deamer [18]. It is instructive to look at their list:

1. An information-carrying polymer must be synthesised by template polymerisation within a compartment delineated by a membrane.
2. Monomers of template synthesis and other raw materials must be provided from the outside, and must be able to pass through the membrane.
3. An external source of chemical energy must be present.
4. Hereditary variation of template replicators must affect the efficiency of catalysed reaction in the protocell.
5. The membrane must be able to grow, either through direct incorporation of membranogenic molecules from the environment or through conversion of appropriate precursors.
6. Reproduction in space of the protocells must happen.
7. Catalysis, replication and growth must be synchronised well.

The chemoton model almost satisfies these criteria: it is less and more at the same time as the minimal cell envisaged by Pohorille and Deamer. Criterion (4) is not satisfied by basic chemotons since templates are assumed to affect the system by means other than sequentially encoded enzymatic catalysis. In contrast, the chemoton has a *bona fide* metabolic subsystem, so it goes beyond criterion (2) in this regard. It is remarkable that the protocell model suggested by the authors includes a pump in the membrane that actively maintains a certain concentration gradient. The issue is analysed in [149]. While a pump may not be strictly necessary, passage of charged molecules through protocell membranes warrants special considerations (see Sect. 2.5).

In a similar vain, Szostak et al. [17] propose that a protocell composed of a growing membrane, a general replicase ribozyme (able to replicate also another copy of itself), and another ribozyme involved at some stage in membrane formation would be truly alive. Once again, it is clear that this system is an "ultimate heterotroph" [19], completely devoid of a metabolic subsystem. We shall come back to this problem when discussing protocellular metabolism.

The aforementioned chemical supersystems we suggest to call *infrabiological* systems: they are not biological, since they always lack one essential component of a minimal living system, but they show a crucial subset of biological phenomena. Among the infrabiological systems those comprising a boundary belong to compartmentalised systems. In this chapter we are going to deal essentially with compartmentalised systems, including systems with chemoton-like organisation. Section 2 concentrates on modes and feasibility of compartmentation, followed by detailed analysis of the nature of plausible genomes of reproducing compartments (Sect. 3). Section 4 deals with the difficult problem of running a metabolism in protocells. We would also like to draw attention to a complementary analysis of related issues by Ruiz-Mirazo and Moreno [149].

2 Compartmentation: Membranes, Reproducing Microspheres and the Lipid World

Without membranes there are no protocells. When discussing their role in early evolution, one should tackle the following issues: formation of mebranogenic molecules, membrane growth and inheritance, microsphere division, and membrane permeability.

It is not our task here to comment on chemical evolution leading to membrane constituents. We rather focus on the remaining issues in turn.

2.1 Autocatalytic Membrane Formation

It is important to point out that membrane growth is an autocatalytic process [20]: membrane constituents are at a high energy state in the aqueous phase; hence they spontaneously insert themselves into pre-existing membrane surfaces. The larger the surface, the faster is the reaction. In this sense pieces of membrane can grow, and if there is some fragmentation, they can multiply. A good question is whether they can have some form of heredity. We mention in passing that in contemporary biological systems there exists the phenomenon of genetic membranes [21]: many biological membranes do not form *de novo*. For example, the membranes of plastids and mitochondria grow by the activity of specific import machineries composed of proteins. These import machineries recognise proteins destined to become parts of the specific cell organelles. The autocatalytic element comes in when we realise that parts of the import machineries are recognised in this way by the import machineries themselves. We are faced here with not only autocatalysis but membrane heredity, as the different kinds of membrane propagate their own kind.

Contemporary membrane heredity rests on the action of proteins, and it achieves only what has been called limited heredity [7, 22]: the number of possible types is limited. Hence evolution is also limited in such systems—this is not to say that they are unimportant. Contemporary heredity is limited because it is based on some subset of shape space, rather than on the whole of sequence space. Any molecule having the right conformation for recognition by the import machinery is dragged in, irrespective of its sequence. (One should not be confused in this discussion by the fact that the proteins involved are genetically encoded. One can imagine the swapping of the import machineries of plastids and mitochondria. This manipulation would immediately redefine membrane identity and propagation without any change in the corresponding genes). Two relevant questions emerge: (i) is it possible to have membrane inheritance without proteins? (ii) What are the limits of heredity in such systems?

2.2 Membrane Growth and Inheritance

An interesting line of research has been initiated by Doron Lancet with his group, conveniently referred to as the "lipid world" scenario [23]. The basic idea is as follows. We know that lipids (more generally: amphiphilic compounds, with a hydrophobic tail and a hydrophilic head) tend to form supramolecular structures, such as bilayers, micelles and vesicles. They can grow autocatalytically. Now imagine that we have a mixture of molecules in any one vesicle. Some of them may act as catalysts of certain reactions. It is theoretically possible that some will catalyse their own incorporation (direct autocatalysis), or that there will be a gang of molecules, each exerting some catalytic function, so that as a net result the incorporation of all members of the gang is ensured by the gang (reflexive autocatalysis). If this idea holds water, membrane heredity in the lipid world, and natural selection of vesicles without a genetic subsystem, would be feasible. The different, reflexively autocatalytic gangs would constitute "compositional genomes" [24]. Note that the model does not deal with the formation of the lipid constituents: they are assumed to be there in the surrounding soup.

Now, there is nothing mysterious about compositional genomes in the first place. Although relying on direct autocatalysis at the molecular level, the genome of the SCM (see section on genetics) is also a compositional genome: the genes there are unlinked, and the genome is characterised by gene composition. Formally, each protocell can be characterised by a genome vector with entries n_i, denoting the number of copies of the ith gene in that vesicle. Change in this number is a stochastic process, which can be characterised by mean and variance.

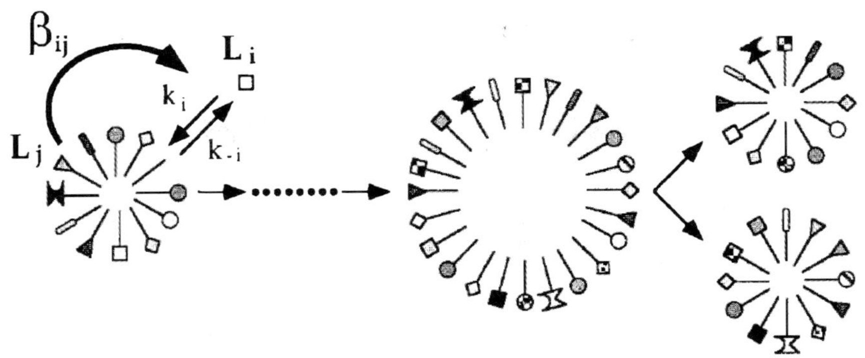

Fig. 3 The GARD model. Catalysed micelle growth and fission. L_i and L_j molecules are different amphiphilic compounds, k_i and k_{-i} are rate constants for spontaneous insertion and emigration of amphiphile L_i, and β_{ij} is the rate enhancement of getting in and out of this molecule from the micelle, catalysed by L_j. Note that the model does not deal with the primary origin of L_i molecules per se (from [23])

A similar approach is possible when considering questions in the lipid world; the issue is however complicated by the fact that we need to tackle the problem of reflexive autocatalysis. This has also precedence in the literature: the reflexively autocatalytic protein networks (e.g. [25]) are perhaps the best known example. We hasten to point out that nobody has seen a real reflexively autocatalytic protein set, apart from very small ones where replication is in fact modular and analogous to the complementary replication of oligonucleotides [26]. Let us see whether one can be more hopeful regarding autocatalytic lipid sets.

The process imagined is shown in Fig. 3. It displays a reflexively autocatalytic micelle with many components. The incorporation of amphiphile L_i may be catalysed by amphiphile L_j at rate enhancement β_{ij} (the ratio of catalysed and uncatalysed reaction rates). The crucial question is this: where can one obtain the values of β_{ij}, considering the fact that no such system has been realised so far (the experimental cases, discussed below, are all directly autocatalytic and show no heredity)? Lancet and his colleagues suggest going around this problem as follows. Fendler and Fendler [27] present a compendium of catalytic reactions documented in micellar systems, from which an empirical distribution of β values can be obtained. Is there a theoretical background to this distribution? The authors suggest translating the model developed for molecular recognition between receptors and ligands [28]. If catalysis depends on recognition of substrate by catalyst, the reasoning is sound: it shows that catalysis is a graded phenomenon. From this empirically constrained theoretical distribution, the authors obtain the β_{ij} values in their GARD model (Fig. 4).

It is instructive to contrast direct autocatalysis with reflexive (mutual) autocatalysis in the GARD model [23]. In direct autocatalysis, the diagonal

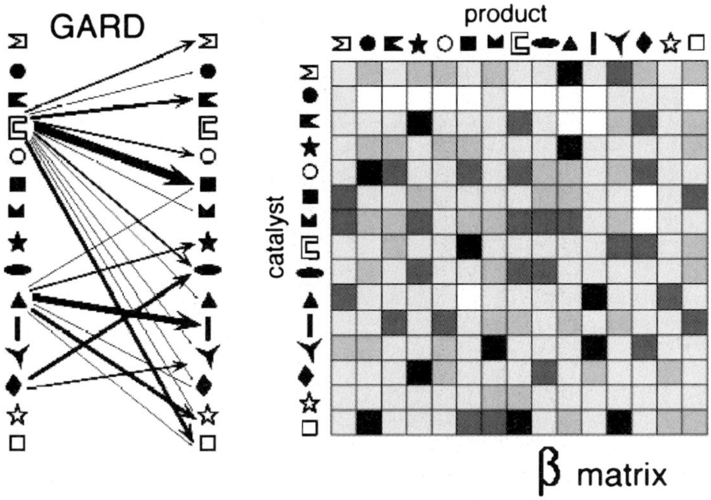

Fig. 4 Visualises example of a β_{ij} matrix generated from a distribution based on receptor/ligand interaction. *Darker blocks* indicate higher numerical values. Note that typically off-diagonal elements are not dominant in this model (from [23])

elements in the catalytic enhancement matrix dominate. Simulations show that, as expected, one such direct replicator will dominate the population. If the diagonal elements are not dominating (as is most likely when using the inferred distribution), then we have the reflexively autocatalytic set.

It is imagined that every micelle (or vesicle) is a *sample* with replacement of a set of possible lipid molecule. Some samples will contain mutually autocatalytic gangs, others will not. The latter ones will not be able to grow. The former will grow and then fragment by some spontaneous process. Micelles containing more efficient gangs (characterised by higher β_{ij} values) will take over. Such sets do have heredity, and they are attractor-based [29]: the gangs maintain and propagate their identity by virtue of their mutual catalytic activity [24].

There are some acknowledged concerns with this model [23]. Reflexively autocatalytic network models (including protein nets [25]) are plagued by the side reaction problem [22]. First, it seems natural to assume that many entries β_{ij} will be effectively zero (no catalytic effect). Worse still, several entries may be negative, which means that in the corresponding reaction an amphiphile catalyses a reaction *draining* the system. The real snag is that in chemical space the vast majority of reactions will be unwanted side reactions. They will presumably be catalysed by values taken from the same distribution. Will this problem demolish the lipid world scenario?

One can say that a similar problem arises in the RNA world: the side reaction problem applies to protocells harbouring ribozymes as well. Not quite, because ribozymes, by virtue of their direct modular replication, can undergo microevolution based on digital information. Genetic information in GARD models is not digital and *small* variations are not heritable (see [30] for a detailed analysis of inheritance in the GARD model).

On the experimental side it is worrying [23] that Fendlers' compendium cites mostly hydrolytic reactions. It may well be the case that different types of reaction (such as biosynthetic ones) should be taken from a different, unknown distribution. This is an exciting problem that should be tackled, irrespective of the lipid world.

2.3 Vesicle Division

The division of protocells is an important issue. In general it should be the result of insertion of membranogenic molecules into the membrane. This can be achieved from outside [31], or from inside. Needless to say, growth from inside is the favoured solution. Whether growth occurs from outside or inside, amphiphiles must be able to jump from one side of the bilayer to the other, otherwise the two layers of the membrane would grow out of proportion. The latter could happen transiently, but not notoriously. Amphiphiles are known to jump from one layer of the membrane to the other (flip-flop mechanism): this reaction must be sufficiently fast in a protocell to render growth and division possible.

In the chemoton model there is stoichiometric coupling between metabolism, template replication and membrane growth, ensuring strict synchrony among these autocatalytic systems. Under such an assumption Rashevsky's [32] early observation on doubling the surface area and volume of a sphere is relevant. Imagine a spherical protocell, kept in that shape by osmotic pressure (turgor) of the metabolites that cannot leave the protocell through the membrane. Due to chemoton coupling, both surface area and the mass of internal material are doubled in growth. But this is incompatible with the maintenance of a spherical shape: surface area scales with the square, volume scales with the cubic of the radius of the sphere. A sphere with double surface area has more than double volume. Since the quantity of osmotically active inner material is only doubled, it follows that the shape of the chemoton becomes continuously distorted. One resolution is that the cross-section becomes dumbbell shaped and ultimately the system divides into two spheres, which together have doubled surface and volume of those of the original [33].

This qualitative reasoning has received confirmation by calculations which assume that the growing microspheres go through a series of quasi-equilibria of different shapes. If the rate of ancient metabolism is not high, this is a fair approximation. Verhás [34] as well as Tarumi and Schwegler [35] were able to calculate the equation of motion for the membrane surface, taking curvature into account. The latter is important since large curvature has higher energy: this is why in general symmetrical shapes are favoured over asymmetric ones.

Božič and Svetina [36] analysed a different situation, where addition of membrane constituents happens from the external milieu, and there is no metabolism inside, but there is limited permeability. They supposed that the membrane assumes spontaneous membrane curvature. This is non-zero if the properties of the inside and outside solutions differ, or if the two layers of a bilayer membrane differ in composition, or if some membrane-embedded constituents are asymmetrically shaped. They were able to show that under these assumptions membrane division is possible provided $TLkC^4 \geq 1.85$, where T is the time taken to double the membrane area, L is the hydraulic permeability of the membrane, k is the bending modulus, and C is the spontaneous membrane curvature. In this model growing vesicles first retain spherical shape, then are distorted to a dumbbell, then to a pair of asymmetric vesicles coupled by a narrow neck, and finally to a pair of spherical vesicles linked by a narrow neck. Separation of the two daughter vesicles occurs as a result of mechanical agitation in the solution.

Heterogeneity of membrane constituents may also play an important role in the stabilisation of vesicles. Amphiphiles with cationic and anionic head groups can assemble into vesicles that are stable over a year [37]. This effect may be explained by assuming an asymmetric distribution of the two constituents between the two layers. Note that the two layers have curvatures of equal magnitude but opposite sign. How such an asymmetric membrane structure would be maintained through generation of protocells is not obvious, however.

It must be said also that temperature can also change the equilibrium shape of membranes because the two layers react to temperature change differently [38], due perhaps to molecular impurities. The obtained shapes are very suggestive. Considering the fact that according to one scenario the origin of replication of longer templates was tied to temperature oscillation [39], it would be important to look at the concomitant changes in membrane shape. Replication as well as membrane division could have been helped by the same oscillation of temperature. Döbereiner et al. [40] demonstrated that budding of sphingomyelin vesicles can be triggered by increasing the area-to-volume ratio of the vesicles by heating. The explanation is that heating induces a gel to sol transition in the membrane. Another way is to induce budding and fission osmotically, which reduces the volume, hence increases the area-to-volume ratio again in the favoured direction. Note that there is no membrane growth in these experiments.

Experiments have confirmed the idea that micelles as well as vesicles could grow autocatalytically (see [41] for a good overview). In a landmark paper Bachmann et al. [42] observed the formation of autocatalytically replicating micelles from sodium caprylate. The micelles could be converted into more stable vesisceles by pH change. Oleic acid/oleate vesicles can also multiply autocatalytically, and they show a remarkable "template" or "matrix" effect in the size distribution: somehow the newly formed vesicles inherit, with "good accuracy" the size distribution of the preexisting vesicles [43, 44]. Microscopic investigation provided evidence for vesicle fission, but whether fission is binary or not is uncertain [45, 46].

2.4 Compartment/Template Infrabiological Systems

Growing membrane systems have been used to obtain artificial infrabiological systems. Walde et al. [47] have carried out the synthesis of polyadenylic acid in self-reproducing vesicles [48], in which the enzyme polynucleotide phosphorylase carried out the synthesis of poly-A, and membrane vesicle multiplication was due to the hydrolysis of externally provided oleic anhydride to oleic acid. The snag is that the enzyme component is not autocatalytic. Enzymatic RNA replication in vesicles [49] suffers from the same problem. It is also not known whether redistribution of the entrapped enzymes into newly formed vesicles occurs or not. An affirmative answer would be evidence for vesicle reproduction by fission.

The demonstration that the polymerase chain reaction can be carried out in liposomes [50] is important because it demonstrates that liposomes can resist the required temperature changes. In the light of the lipid world model it is useful to ask what catalytic functions Luisi's structures show behind direct autocatalysis. Binding of peptides to and polymerisation of amino acids in liposomes was demonstrated in various systems [51]. We are not aware of a similar effect on nucleic acid synthesis.

The direct autocatalytic multiplication of both caprylate and oleic acid vesicles received a simplified kinetic analysis [52]. It was shown, in agreement with elementary reasoning that the catalytic effect is due to large growth of the reaction surface.

Another line of research demonstrated the catalysis of membrane formation from micelles by montmorillonite surfaces [53]. Such vesicles grow in size but spontaneous division (without externally enforced extrusion) was not demonstrated. RNA absorbed to clay can be taken up together with clay particles, but neither clay nor RNA is autocatalytic in this experimental system. Fatty acid vesicles containing RNA have a higher internal osmotic concentration. This is expected to strain the vesicle membrane and facilitate its growth. RNA-containing vesicles are shown to grow in size at the expense of vesicles that do not contain RNA [54]. The process is likely due to fatty acid molecules leaving the isotonic vesicles (which hence shrink in size) and joining the RNA-containing vesicles (which hence increase in size). The process continues until a new equilibrium is reached. Competition between protocells is thus an intriguing possibility not yet experimentally demonstrated.

In an attempt to design a protocell, a Los Alamos group proposed a system essentially composed of non-enzymatic template replication coupled to micelle growth [55, 56]. The micelle aggregate is assumed to incorporate from the medium precursors of lipids and template building blocks (monomers or oligomers). The authors assume that for this particular construct PNAs [57] would serve better because of their hydrophobic nature. It is assumed that single-stranded molecules face the hydrophilic anterior whereas double-stranded molecules immerse into the hydrophobic interior of the micelles. Alternation between these two states is assumed to facilitate replication.

A possible form of coupling between membrane growth and template replication is assumed to proceed as follows. PNA molecules are planned to carry photosensitiser molecules. PNA acts as a catalyst for the light-driven fragmentation of the lipid and template monomer precursors. It is also assumed that certain PNA sequences provide the efficient charge transfer system necessary to realise the dynamic coupling.

It remains to be seen whether this clever proposal can be chemically realised. If yes, it would serve as a proof-of-principle, but two main limitations remain. Due to the envisaged template replication, template length is assumed to be less then about 10 [56]. It seems that the growth dynamics, essentially obeying the "survival of everybody" [58] rule for parabolic replicators [150, 151], would result in a diversity of short oligomers, but the situation is more complicated. As mentioned above, only certain sequences would act as efficient charge transfer systems. Thus selection for efficient growth would result in very few winning, short sequences. Presumably, one would arrive at a system with limited heredity [7] and evolution would not be open-ended. It is doubtful whether any functionality of PNAs other than charge transfer could evolve in the system. It is not easy to see how the Los Alamos Bug could be modified to become a unit of evolution.

2.5 Membrane Permeability

Membrane permeability is one of the most difficult problems for protocells to solve. Electrolytes, amino acids and sugars permeate liposomes at a very small rate. Today we find highly evolved transport mechanisms, resting on evolved proteins, which solve this problem for the cell. Protocells must have resorted to more basic, but sufficiently functional tricks.

A number of suggestions have been made. A simple mechanism was proposed by Stillwell [59], resting on the following components: (i) increase the permeability of a molecule by a transient chemical modification, which renders the molecule more lipophilic; (ii) maintain a gradient by lowering the concentration of the material in question by metabolising it (Fig. 5). In concrete terms, amino acids [59] and sugars [60] could react with alde-

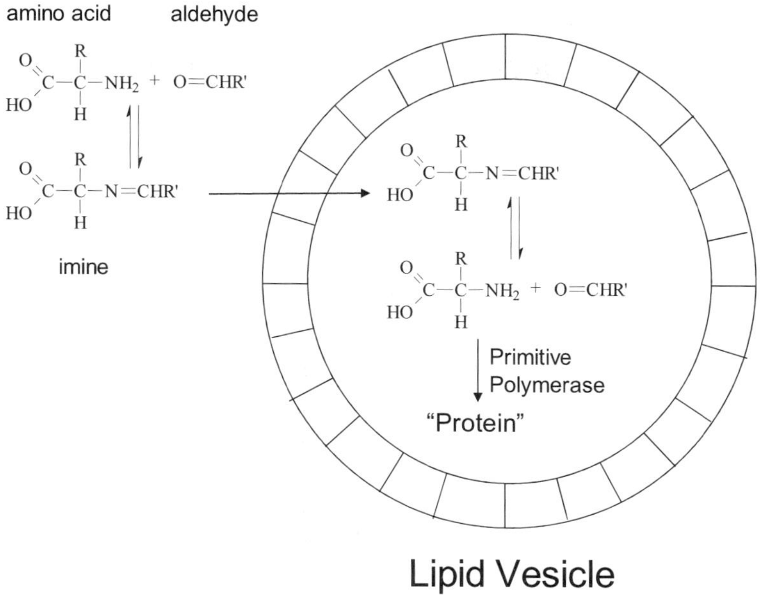

Fig. 5 Enhancement of amino acid transport into vesicles by reversible chemical transformations. Amino acid reacts with aldehyde to form a more permeant molecule. Inside the vesicle the reaction runs in reverse and the amino acid can be incorporated into other molecules, such as peptides

hydes (such as formaldehyde and pyridoxal) to form a so-called Schiff base, which then could pass more readily through the lipid bilayer. Experiments have confirmed the feasibility of this idea. Another possible route, requiring a H^+ gradient across the bilayer (acidic inside), was suggested by Chakrabarti and Deamer [61], pertaining to amino acids. The most common form of amino acids is zwitterionic, and the less common form is the neutral (uncharged) molecule (Fig. 6). The proposed transport mechanism consists of the following steps: (i) neutralise the $-COO^-$ group by creating amides or methyl ester derivatives; (ii) at basic pH the $-NH_3{}^+$ groups lose their H^+; (iii) let the neutral molecule pass through the bilayer; (iv) acidic pH inside restores $-NH_3{}^+$ and renders the amino acid again impermeable; (v) use up the amino acid in metabolic reactions. The increased transport rate of amino acids thus neutralised has also been experimentally demonstrated.

An interesting mechanism for the establishment of a pH gradient in growing fatty acid vesicles was recently shown by Chen and Szostak [62]. Fatty acid vesicles are usually very permeable to cations, including H^+. Maintenance of a gradient thus requires a non-permeant cation, such as arginine. Incorporation of protonated (neutralised) fatty acid molecules from the *external* medium results in acidification of the internal milieu, by a flip-flop mechan-

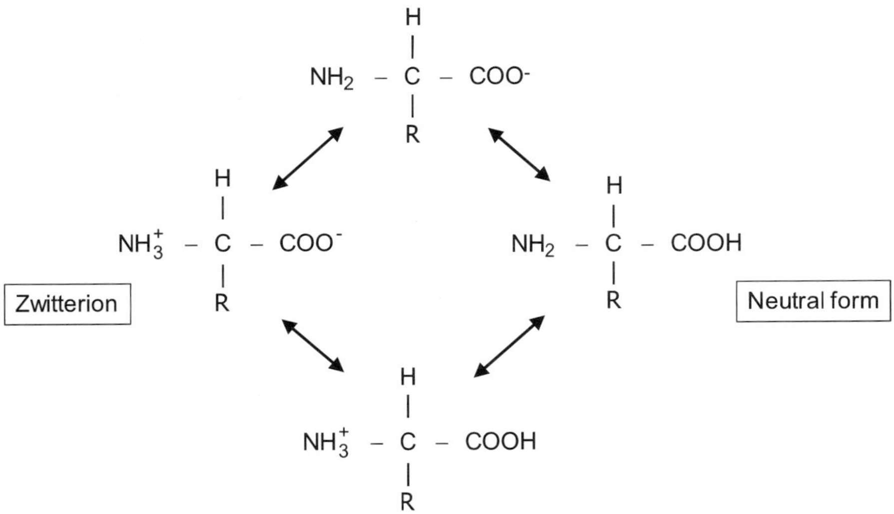

Fig. 6 Another mechanism of amino acid transport rests on the idea of the reversible formation of the neutral form, which passes through the membrane much faster

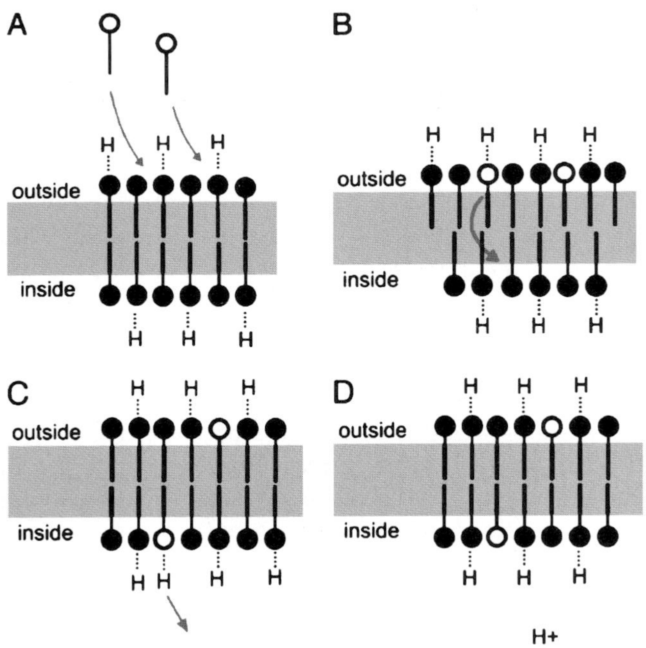

Fig. 7 Membrane growth (**A**) can create a pH gradient. At the pK value of the fatty acid head of the incoming fatty acid molecules will be protonated (**B**), hence neutralised. Neutralised molecules flip more readily (**B**). Approximately half of the flipped molecule will be deprotonated (**C**). A pH gradient (**D**) builds up provided the membrane is rather impenetrable to other cations (from [62])

ism of H^+ transfer, performed by the fatty acid molecules (Fig. 7). The energy released in membrane growth can thus be stored in the pH gradient thus built up. Such a gradient could be coupled to some other processes, as discussed above.

3 Primordial Genomes

The genome is the total genetic constitution of an organism. In present-day cells information encoded in DNA is transcribed into RNA and translated into proteins, the absolutely essential polymers for almost all cellular functions. But such a flow of genetic information and relatively clear-cut boundary between replicating and catalytic systems was a later invention. Primitive cells (protocells) may have been entirely RNA-based, which circumvents the historical dilemma of which came first, proteins or DNA. There was little separation between molecules acting as functional units and those acting as genetic material. Nevertheless, even if difficulties related to prebiotic RNA synthesis and stability can be eventually solved, an important problem still remains: How could information have been preserved when templates were confined to isolated vesicles of finite size and replicated by a low copying fidelity RNA polymerase ribozyme?

3.1 Error Thresholds

The pioneering works of Eigen [63] called attention to the fact that the length of selectively maintained genetic information is limited by the copying fidelity. Without the aid of peptide enzymes the upper bound of copying fidelity per nucleotide per replication was likely around 96–99% [64, 65]. With such a high mutation rate the number of mutated offspring molecules far exceeded the number of non-mutated ones, resulting in the well-known quasispecies concept of Eigen [7, 63, 66–68] where a stable cloud of mutants formed around a master sequence as far as the maximum chain length (N) was below the critical copying fidelity per site per replication $(q*)$ as determined by the following simplified expression:

$$N < \ln s/(1 - q^*) ,$$

where s is the superiority of the master. The occurrence of thresholds for error propagation was originally derived as a deterministic kinetic theory that is only valid in the limited case of an infinite number of molecules. Alves and Fontanari [69] have extended this to finite populations and found that q^* increases linearly with the inverse of population size.

3.2 Vesicle Models

Proposals to circumvent the information crisis have focused on networks of non-encapsulated cooperative molecules (i.e. hypercycles, Fig. 8), or compartmentalisation of competitive unlinked templates (i.e. vesicle models, Fig. 9). As originally formulated [63, 66] the hypercycle is a catalytic feedback network in which each template helps in the replication of the next one, in a secondary cycle closing on itself (second-order autocatalysis). The enzymatic function (replicase activity) in the hypercycle was thought to be carried out by the encoded proteins (the "realistic" hypercycle [70]), but recently also the hypercycle was projected into the RNA world [8], with RNA molecules acting as templates as well as enzymes [71]. Problems arise, however, due to the dynamical instability of large hypercycles (e.g. they are vulnerable to extinction via fluctuations [72]), and the evolutionary conflicts among its members no matter how small hypercycles can be. Spatial implementations of hypercycles simulated by cellular automata are of little aid in this matter as it was shown [13] that the spatial patterns (spirals) that would supposedly increase the robustness of hypercycles against parasites [73] are unstable against a trivial form of "patchy environment", namely differential death rates of replicators in the cells of the grid. As first recognised by Maynard Smith [74], to provide catalytic support in a molecular catalytic feedback network is an altruistic behaviour doomed to exploitation by parasitic molecules and eventual extinction. Apparently this criticism still holds.

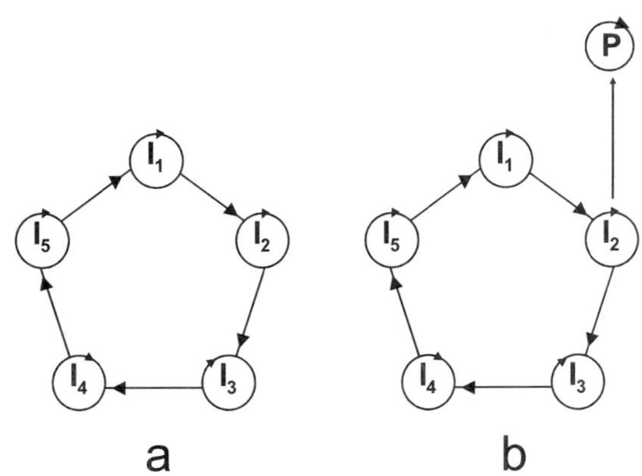

Fig. 8 The hypercycle (**a**) and its parasite (**b**). Each member I_i is autocatalytic for its own growth and heterocatalytic for the replication of the next member. The parasite P shown accepts the catalytic help from I_2 but does not give anything back. If the arrow leading to P is stronger than that leading to I_3, the system is doomed to extinction in a spatially homogeneous dynamical system

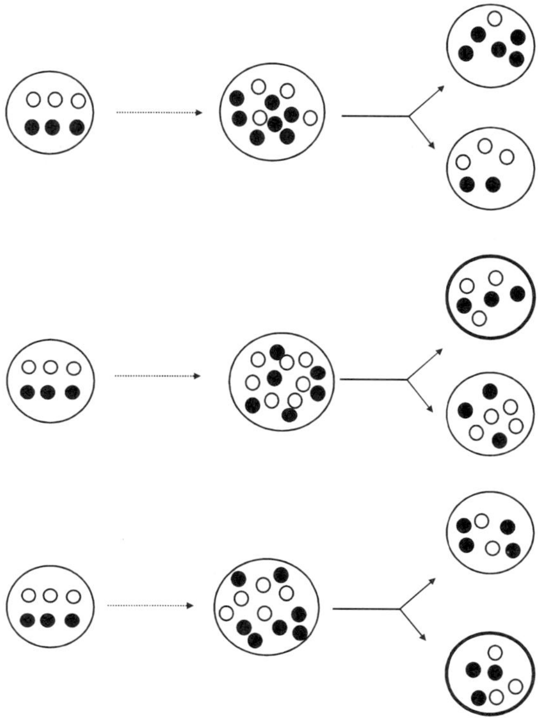

Fig. 9 One of the vesicle models (as depicted by the SCM). Different templates (labelled by *open* and *closed circles*) contribute to the well being of the compartments (protocells) in that they catalyse steps of metabolism, for example. During protocell growth (--→) templates replicate at differential expected rates, but stochastically. Upon division (→) there is chance assortment of templates into offspring compartments. Stochastic replication and reassortment generate variation among protocells, on which natural selection at the compartment level can act and oppose to (correct) internal deterioration due to within-cell competition

The initial stimulus behind vesicle (compartment, protocell) models was to solve the conundrum of the evolutionarily dynamic coexistence of unlinked genes in view of the problems faced by hypercycles. Sooner or later hypercycles would have had to escape into compartments (as already accepted by Eigen et al. [75]), but in those circumstances alternative systems such as a group of competing species of molecules (genes) that happened to be enclosed in a vesicle and were replicated by a non-specific replicase could have been a more efficient alternative for accumulating information and for forming a catalytic network that otherwise could be unstable in homogeneous solution [17, 19, 76–79]. Elaborating on the package model proposed by Niesert et al. [80], Szathmáry and Demeter [81] described the SCM as a dynamic system where an optimal template composition ensures the fastest growth and division of protocells. The protocell grows due to replication of templates and by adding the necessary membrane building blocks due to the metabolism of the compartment. The fissional size corresponds to a certain level of polyploidy of the genes (i.e., gene redundancy). As genes are distributed randomly to the daughter cells dead offspring can arise, and their average fraction increases with decreasing polyploidy of the slowest replicating template. For this reason the initial formulation of the SCM assumed that the number of different template types per protocell must have been small.

A common criticism faced by the SCM is that it is a very sloppy system of information integration because the major difficulty for conserving a complete set of genes is the growth rate difference between replicators [80, 82, 83]. The snag with such criticism is that it overlooks evolution. Thus, if we reasonably assume an evolving population of protocells where there is initially genetic variation for replication rates, only those lineages with a reduced variance in growth rates among unlinked genes would eventually survive [84]. Those lineages could likely host many more genes than the alleged upper limit of 3 (e.g. [80, 82]). Actually, in the limiting case of an infinite number of vesicles, each one carrying a finite number of templates that replicate at the same rate, Fontanari et al. [85] have recently found that there is no fundamental impediment to the coexistence of an arbitrary number of template types inside a vesicle (except of course for the vesicle capacity). This is in sharp contrast with the inherent dynamical instability of large un-packed hypercycles (see above). Since the total information content is the product of the number of different templates and the maximum information coded per template, the information gain due to the coexistence of different templates in stable hypercycles relative to the theoretical upper limit in vesicle models is therefore negligible.

Compartmentalisation thus offers the most natural and efficient way of information integration, but it does not still solve the original problem raised by Eigen [63]: the information crisis in primitive genomes. The first attempt to compare q^* between "conceptually analogous" versions of compartmentalised hypercycles and the SCM was carried out by Zintzaras et al. [71]. They found that a population of SCM protocells can tolerate a higher input of deleterious mutation rates, and reaches an equilibrium mutational load (i.e. the decrease in average fitness of a population exposed to deleterious mutations relative to an error-free population) lower than that in a population of protocells hosting hypercycles. Hence, given our current understanding a working-model for the protocell scenario in the origin of life would enclose a persistent cloud of mutants around master sequences of genes competing for within-group common resources. Selection on stochastically produced offspring variants would favour those lineages with non-competitive molecular assemblies, thus stabilising the population against random loss of essential genes after compartment fission. But a critical question still remains: Could such a primitive protocell sustain the "minimum" informational length required for the basic features of life with a putative low copying fidelity RNA polymerase ribozyme?

3.3
Recombination (Sex) and Gene Redundancy

Within the classical framework of vesicle models, the previous question was approached by several authors [71, 86] and the answer was clear: compartmentalisation per se was not sufficient to overcome the information bottleneck imposed by the error threshold. However, Lehman [87] raised the issue that recombination—a frequently ignored player in models of early evolution—could have been crucial to build up primeval genomes of sizeable length. In the article that coined the phrase "the RNA world" Gilbert [8] already speculated that "the RNA molecules evolve in self-replicating patterns, using recombination and mutation to explore new functions and to adapt to new niches". In this context the discovery of spontaneous rearrangements and recombinations of sequence-non-specific RNAs in solution is important [88]. According to the experimental evidence, RNA chains of diverse sequences can recombine at a rate of $10^{-9}\,h^{-1}$ per site and the reaction is not due to cryptic ribozyme structures that might be formed by some RNAs, but is an intrinsic chemical property of polyribonucleotides. More recently, Riley and Lehman [89] have shown that *Tetrahymena* and *Azoarcus* ribozymes can promote RNA recombination.

This capability of RNA to potentially minimise the burden imposed by the error threshold—along with [87]—has been recently analysed by Santos et al. [90]. They assumed that recombination in protocells took place via copy-choice means; i.e. that the replicase switched between RNA-like templates as occurs frequently in RNA viruses and is crucial for retroviral replication during reverse transcription (e.g. [91–95]. They ignored, however, the possibility of gene chimerisation resulting from illegitimate recombination—a putative source of major evolutionary innovations as discussed by Cavalier-Smith [96]—because it would introduce many technical and theoretical problems in a protocell scenario. The numerical results showed that there is a quite intricate interplay between mutation, recombination and gene redundancy, but the conclusion from the fitness function they used was that the informational content could have increased by $\sim 25\%$ at most by keeping the same mutational load as that for a population without recombination. Even so, the upper bound of ~ 75 nucleotides reached in that work is still far from the minimal life provisions.

The consequences of imperfect replication in vesicle models are somewhat puzzling [85, 90]. For small mutation rates an increased level of polyploidy favours the persistence of protocell lineages since the random loss of essential genes after fission is attenuated. However, for large mutation rates the situation is reversed, resulting in that those lineages with low levels of polyploidy are better able to cope with higher mutation rates, particularly when recombination is allowed. This means that gene redundancy was indeed costly. Therefore, selective forces favouring the linkage of genes to make the first chromosomes would eventually outweigh the advantage of faster replicating single genes because linked genes are less likely to be lost by random assortment when protocells divide [97].

The role of the number of gene copies in a primitive cell was investigated by Koch [98], who pointed out the existence of two conflicting forces: (i) higher copy numbers act as a safeguard against random loss of all copies of a gene; (ii) but such copy numbers slow down adaptive evolution because a newly arisen favourable mutant is diluted out and cannot be "seen" efficiently by natural selection acting on cells. He further observed that a moderately high (< 100) copy number per gene is not only optimal, but is confers some additional evolvability by the "duplication and divergence" scenario, as first emphasised by Ohno [99].

3.4
Lessons from "Bags of Genes" in Contemporary Genetic Systems

It is important to pint out that unlinked, independently replicating and reassorting genetic elements (replicons), exist even today. For example, the macronucleus of ciliates (unicellular protists) is essentially a bag of genes in high copy number [100]. Bacteria can harbour a number of chromosomes and plasmids. It is important that high- and low-copy-number plasmids follow different assortment strategies: the former rely on stochastic assortment, whereas the latter apply a cell wall-mediated accurate segregation mechanism [101], in line with conclusions on analogous protobiological systems. A further remarkable example is that of the unigenic mini-circles of the dinoflagellate plastid [102, 103]. It is important to stress that all these systems are only partially analogous, rather than homologous to the assumed primordial "bag-of-genes" genomes.

It is potentially rewarding to have a closer look at the ciliate case. Ciliates usually have one or a few micronuclei and one or many macronuclei (Fig. 10). It is only the former that participate in sexual recombination; we are not aware of a case of macronuclear fusion. Typically, a round of sex is followed by hundreds of clonal cell divisions (called the vegetative phase). During vegetative reproduction the micronucleus is (almost) completely inactive and the macronucleus serves the transcriptional needs of cells. DNA in the macronucleus is highly amplified, presumably to meet the transcriptional demand of the large ciliated cells (they can be larger than 100 micron in length). DNA in the macronucleus originates from one of the micronuclei after fragmentation, elimination and specific amplification [100]. *Paramecium* and *Tethrahymena* species show these phenomena to a moderate extent: In the latter, the macronucleus contains about 57 copies of the unique se-

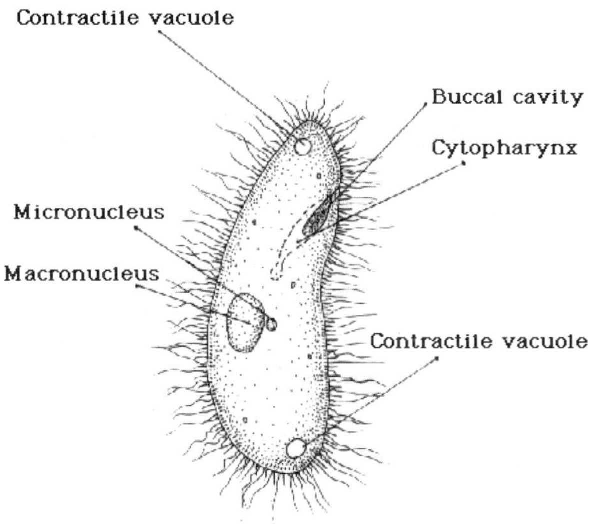

Fig. 10 Micrograph of the ciliate *Paramecium*. Macronuclei of ciliates are more or less fragmented genomes (bags of chromosomes or genes) (Source of the image: http://www.bio.umass.edu/biology/conn.river/parameci.html)

quences found in the micronucleus. Out of the 5 chromosomes up to 200 different subchromosomal DNA molecules are generated. Thus each such fragment still contains thousands of genes.

Macronuclear reorganisation is much more radical in the co-called hypotrich ciliates, in which fragmentation occurs down to the level of the gene. Finally, each unigenic minichromosome is endowed with the sequences necessary for replication. For example, in *Oxytricha*, the macronucleus harbours about 24 000 different unlinked genes, each with an average copy number of 950.

The odd fact is that upon division bulk DNA material, as well as each copy of genes, is *randomly assorted* into offspring nuclei. Thus there can be 10–15% difference in the total DNA content of sister macronuclei. This calls for some correcting mechanism at the genomic as well as the genic levels. For the former, chromatin extrusion, extra full or partial rounds of DNA replication, and/or skipped rounds of replication all seem to contribute to varying extent, although the molecular basis for DNA content regulation is unknown.

At the genic level we find that different genes can be maintained at various copy numbers [104]. This set number can change abruptly, presumably because of mutation in the regulatory sequences [100]. It seems that the added telomeres at the ends of the minichromosomes are sufficient for the initiation of replication [105]: chromosome-internal sequences do not seem to influence the regulation. This of course renders the puzzle of copy number control *specific* for different (groups of) genes even greater.

We wish to draw some important conclusions from the ciliate case. Apparently, the strategy of the bag-of-genes is a viable one, and is a satisfactory base of cellular inheritance. (Note that the objection that macronuclei are doomed to death and rejuvenation from sexually recombining micronuclei is mandatory is flawed—there are known cases of amicronucleate clones that can be maintained apparently indefinitely [100]). This does not resort to maintenance of linkage groups or accurate segregation, but does require some mechanism for suppression of replication and copy number control. It seems that the basic adaptation to ensure that most molecules are replicated exactly once in hypotrichs is the existence of so-called replication bands, in which individual DNA molecules cannot migrate and there is a directional wave of DNA replication [100]. Nevertheless, inaccurate segregation requires an active correction mechanism for copy number, as discussed above.

The hypotrich ciliate is very suggestive for early evolution studies. First, it should not escape our attention that one suggested mechanism for initiating replication of RNA molecules is by telomers; noting that telomerases have a ribozyme component and that some RNA viruses adopt a similar strategy [106]. Ciliates testify that a very large number of genes can be maintained when copy number control sufficiently reduces the assortment load. It would be very important to find out the molecular details of the ciliate mechanism, since some analogous mechanism could readily ameliorate the earliest genomic conflicts. The apparent independence from chromosome-internal sequences is particularly encouraging.

Multicopy plasmids offer a limited, but still instructive comparison. Plasmids are replicating clusters of genes in bacteria. Under certain conditions they confer beneficial traits (such as resistance against antibiotics) on the bacterial host. They can also be passed on by conjugation between cells. A well elucidated mechanism of copy number control of abundant plasmids relies on a *trans*-acting inhibitor and *cis*-acting activator. For example, the ColE 1 plasmid in *Escherichia coli* blocks the action of the RNA primer (activator) by

Fig. 11 Sex between protocells. The model assumes that protocells host essential replicators (genes) with metabolic function that are replicated by non-specific replicases (R). According to the replication rate constants the replicase can be "selfish" (R1; i.e. reaps the benefits of a common metabolism and swiftly outgrows the metabolic genes), "cooperative" (R2; i.e. all genes inside a vesicle grow at nearly similar rates) or "altruistic" (R3; i.e. helps the metabolic genes but at the expense of deviating the whole compartment from the optimal gene composition). The figure plots some sample simulations showing the average number of different replicases per protocell according to the proportion of cells that undergo random cellular fusion (sex). R3 is quickly lost, but the frequencies of both R1 (*grey lines*) and R2 (*black lines*) oscillate according to the "amount of sex" (proportional of cell fissions/generation). The important point here is that a protocell population could resist invasion of rapid exploitation by a potentially lethal (at the compartment level) parasite (R1 in this case) (After [84]) ▶

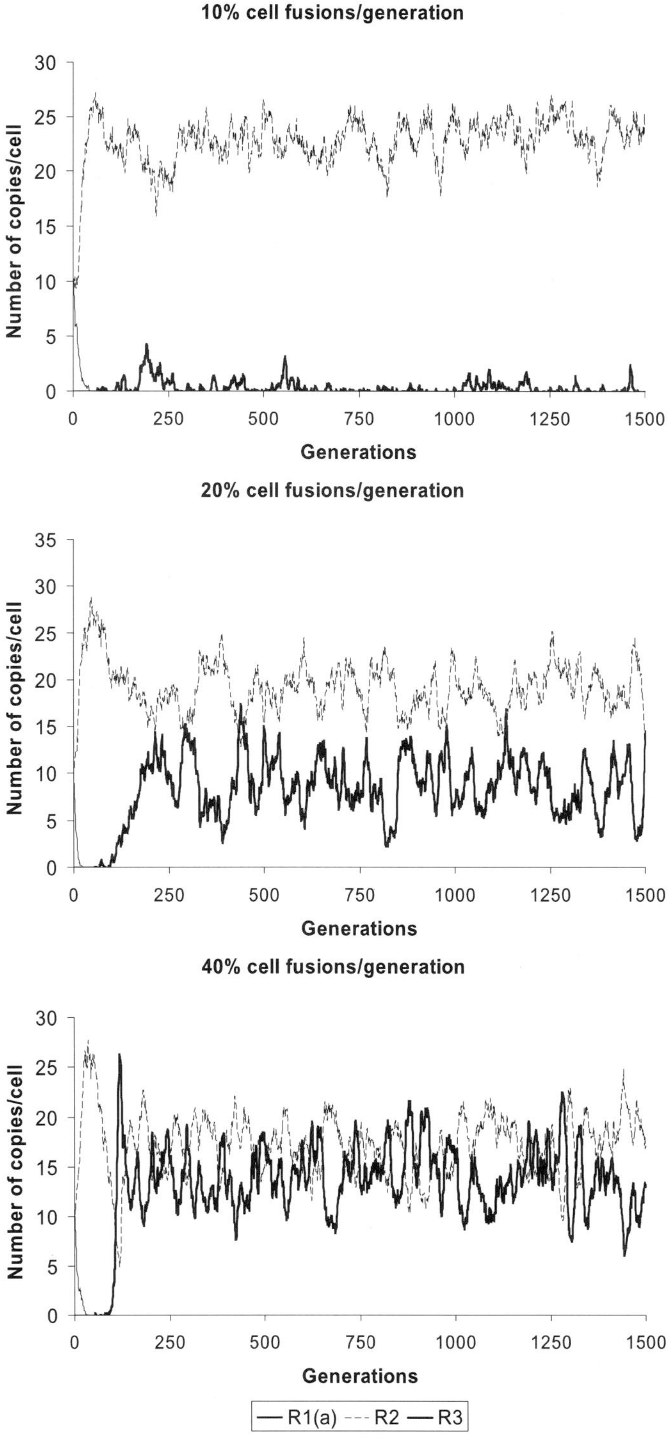
10% cell fusions/generation
Number of copies/cell
30
25
20
15
10
5
0
0
250
500
750
1000
1250
1500
Generations
20% cell fusions/generation
Number of copies/cell
35
30
25
20
15
10
5
0
0
250
500
750
1000
1250
1500
Generations
40% cell fusions/generation
Number of copies/cell
30
25
20
15
10
5
0
0
250
500
750
1000
1250
1500
Generations
R1(a)
R2
R3

a constitutively produced, unstable inhibitor RNA. The concentration of the latter is a good indicator of plasmid copy number in the bacterium [101]. It is true that several unrelated plasmids can coexist in the same bacterium, but closely related plasmids cannot be stably maintained in the same population, unless reintroduced by horizontal transfer. This reminds one of the facts that ciliate macronuclei usually become homoallelic because of chance segregation of the alleles of any given gene; but there is a crucial difference. The latter process happens even if the alleles in the same cell do not compete. In contrast, closely related plasmids vary in the strength of activation and inhibition elements, and there is therefore within-cell competition [107]. Between-cell competition is influenced by at least three factors: (i) whether the plasmid is essential for survival in the given medium; (ii) very high copy number entails a metabolic burden; (iii) horizontal spread of plasmids is possible.

A final remark here is that in vesicle models the classical source of intragenomic conflict (i.e. conflicts among different elements of the genome) is thought to arise because natural selection acts at two levels: within- (i.e. selfish replicators can reap the benefits of a common metabolism to enhance their own survival) and between-protocells. Sex (broadly defined as the exchange of genetic material between genomes or between two sources [87, 108]) poses yet another riddle because genetic systems that involve fusion between organisms (protocells) offer higher prospects for parasitic genes. The menace of horizontal transfer of parasites between protocells was studied by Santos et al. [84]. They concluded that a population of protocells is able to resist invasion of parasites and, in some situations (i.e. when an over-exploiting parasite invades the population), extensive cellular fusion would have been beneficial and the argument by Hamilton et al. [109] for sex as an adaptation to parasites applies (Fig. 11). An important point is that the scenario they explored numerically is fully consistent with the idea that life may have begun as a series of ever-changing, swapping committees of proto-organisms that exchanged much genetic information [110, 111]. What remains to be explored is the flip side of the coin: the potentially important role of co-opting genetic material from other protocells to significantly speed up evolution.

3.5
Minimal Set of Genes for Cellular Life: The Top-Down Approach

Comparative genomics shows that most bacterial proteins are highly conserved in evolution. From this standpoint an increasingly appealing issue is the identification of "the smallest possible group of genes that would be sufficient to sustain a functioning cellular life form under the most favourable conditions imaginable, that is, in the presence of a full complement of essential nutrients and in the absence of environmental stress" [112]; in other words, to define "the minimal gene set". The latest suggestion by Gil et al. [113] is a minimal gene set composed of 206 genes. It is highly illustrative to list some of the molecular features apparently needed for the hypothetical simplest bacterial cell:

1. A virtually complete DNA replication machinery.
2. A rudimentary system for RNA repair.
3. A complete transcriptional machinery.
4. A nearly complete translational system.
5. Protein-processing, -folding, secretion and degradation functions.
6. Machinery for cell division.
7. A basic substrate transport machinery.
8. etc. ...

From this catalogue it is painfully obvious that a great deal of molecular and protocellular evolution preceded the hypothetical "minimal cell" in the context of comparative genomics. Our point here is that the top-down approach to design a minimum cell in terms of molecular biology is a worthwhile exercise, but logically and evolutionarily comes later, and not instead of, the chemoton. In addition, the comparative genomic approach might be flawed for the following reason. Suppose that genomes consist of four genes: A, B, C and D in bacterium B1, and A′, B′, E and F in bacterium B2. Genes C, D, E and F show no homology whatsoever. This does not mean that (A, B) and (A′, B′) organisms would be viable. They may have a problem, which must be solved somehow, and it happens to be solved by genes D and C in B1 and genes E and F in B2.

4 Metabolism

4.1 All Living Systems Today are Metabolic

All living systems contain metabolism consisting of at least one autocatalytic cycle. An autocatalytic cycle is a set of consecutive reaction steps that has A_i constituents ($i > 1$), takes as material inputs a set of reagents, X, and produces a set of products Y. A cycle has an autocatalytic stoichiometry if, after a finite number of turns, each constituent multiplies in quantity [5]. In the simplified description of the formose reaction (Fig. 12), in one turn of the cycle, two molecules of glycolaldehyde are formed for each glycolaldehyde molecule present at the beginning. Therefore, it has autocatalytic stoichiometry. The products, Y, are used in constructing the organism (e.g. membrane, templates), or expelled as waste materials from the chemoton. Chemical cycles are homogeneous catalysts. Catalysts accelerate chemical reactions without being used up or changed in nature. They act at low concentrations by reduc-

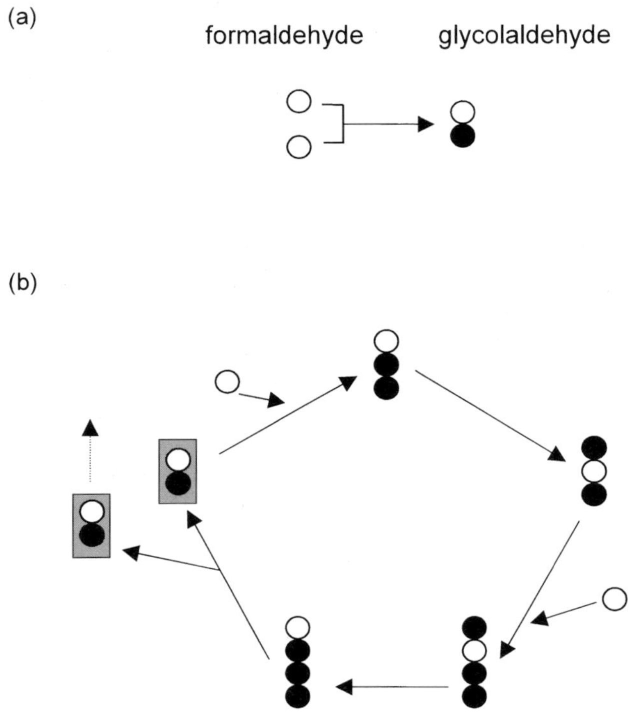

Fig. 12 The formose contains an autocatalytic core, in which the amount of cycle intermediates (such as glycolaldehyde) grows autocatalytically

ing the activation energy of the reaction. All living systems are catalysts, e.g. yeast catalyses the production of CO_2 from sugars, but in addition they use some of the chemical matter of the reagents to construct their constituents hence they are autocatalytic.

Why should metabolism be an absolute characteristic of life? An organism without metabolism would be one that did not synthesise any of its materials from precursors, but obtained them all pre-formed from the environment and from its parent(s). Such an entity could only exist if all its materials were present in the environment in sufficient concentrations. Heterotrophic theories popularised by Oparin [114] and Haldane [115] assume such concentrations could have been found in a rich prebiotic soup. Lancet's GARD model [23, 24] and other models of reflexive autocatalytic sets such as Eigen's hypercycle [66], Farmer's et al. [116] autocatalytic binary strings, Fox's microspheres [117], and more recently Szostak's protocell [17] all make the same assumption. How much time would it take the biosphere to deplete this free gift of complex molecules? This would depend on the "gross primary production" of the primitive biota [118]. Although the hypothesised entities do not metabolise X in order to synthesise their components, any complex organisation requires some energy for self-construction, i.e. to reduce internal entropy. Unless all this energy could be obtained from sources external to X (e.g. light or redox potentials), X would have to be used as a chemical energy source, and so degraded to waste products. Also X decays to simpler molecules at a finite rate. This happens independently of any reactions required for the maintenance of the organism. Therefore, long-term persistence of non-metabolising entities assumes the existence of mechanisms in the environment able to replenish X. No known mechanisms, other than living systems with metabolism, are capable of synthesising complex organic molecules continuously. Since there is no continued influx of complex organics from outer space, non-metabolising organisms can therefore only exist as transients in a system initialised with abundant complex organic molecules, because eventually these will run out. Dyson suggested that life may indeed have started in this way [119]. Alternatively, non-metabolisers can be "parasitic" upon complex organics produced by entities capable of synthesising complex organic molecules from a subset of X, as are viruses for example. In the long-term the concentration of these metabolising entities will be the rate-limiting factor for the non-metabolising entities. So metabolism should be considered a characteristic of living systems because understanding metabolism is crucial for explaining how living systems work as dissipative structures [120].

It cannot be stressed too strongly that without exception, all known cellular life possesses an autocatalytic metabolism, even if the cells are heterotrophic. Thus for the autocatalytic nature of the whole metabolic network it is not necessary to be able to identify a smaller autocatalytic core as the reductive citric acid cycle or the Calvin cycle. Imagine the following thought experiment. Take away all metabolites from a cell but leave all the water and the informational macromolecules in place. Can the network be recreated from the food materials only, or not? Let us be generous and provide enough ATP also for the supposed kick-start. The fact is that no contemporary cell could resume its activity in this experiment. Consequently, all cells today possess a *distributive* autocatalytic network that *cannot be seeded from outside*, because some of its seed components cannot be taken up from medium.

It is easy to see that such a system has certain advantages in bad times, when food is depleted. Because some key components of the network cannot permeate the membrane, they cannot leak out either, thus a package of metabolites will be preserved in the cell interior (until it is degraded by side reactions). In contrast, a cell whose metabolic network could be re-created entirely from outside could also lose all it metabolites under harsh conditions since they would simply be lost by reactions running in the reverse. We suggest the term *endogenous autocatalysis* to describe contemporary metabolic networks.

Living systems exist in a biosphere that to a first approximation is a closed thermodynamic system (i.e., conserves matter but not energy, like a greenhouse). To explain life's persistence, we must explain how a finite set of chemicals can be recycled effectively and indefinitely by a biosphere. As a prerequisite for the persistence of life, we require entities that are capable of obtaining energy from outside the system in order to re-cycle the chemical system (autotrophs or non-living "autotrophic" metabolic systems). Entities capable of recycling a subset (X1) of chemicals using only energy from another subset (X2) of chemicals will not be able to do so for very long, since X2 will run out. Heterotrophs are an example of such entities. If they preceded autotrophs historically, then we can conclude that they could only persist if either autotrophs or non-living autotrophic metabolic systems could evolve before X2 was used up, killing all the heterotrophs. Let us consider the requirements for the evolution of non-living metabolic entities of both "heterotrophic" and "autotrophic" types.

4.2 Evolution of Autocatalytic Cycles

Imagine an experiment that simulates early earth conditions [121, 122]. Construct multiple "micro-environments" each with different characteristics. These could use different abiotic energy sources, for example. Some might use UV light (oscillating as night and day) or redox potentials (e.g. FeS_2/FeS surfaces). They could also have different temperatures, salinity, pH, local chemical concentrations or other attributes. Together, these disparities will make the equilibrium positions and chemical reaction rates vary between these micro-environments. The chemical network is massively reconfigurable and non-linear. Spatial factors can establish chemical gradients, so some degree of specificity over and above that provided by the chemical network can be obtained. In this way, spatial properties of the chemical network enable "vectorial metabolism" [123]. In essence, we can keep the system very far from equilibrium, in many different ways, physically and chemically. Initialise the system with a subset of atoms and small molecules e.g. C, H, N, O, P, S, and leave it for some time. *Under what circumstances will the system settle down into a boring point attractor, e.g. tar, and under what circumstances will it produce life?* Would the "tar" be the same boring tar, or different and still boring tar, if the "tape were re-run"? Under what circumstances would an autocatalytic cycle arise? And for what subset of parameters would an autocatalytic cycle evolve into life? How would the "platonic space" of all possible chemical reactions be explored [122]?

King [124] modelled a recycling chemical network (i.e., where every molecule type is produced in at least one reaction, and consumed in at least one reaction) of bimolecular reactions (i.e., where two reagent molecules react to produce two product molecules) and showed that the number of "platonic" autocatalytic cycles C is given by

$$C = \sum_{i=0}^{r} J_i - r,$$

where J_i is the number of reactions that the *ith* reactant takes part in, and r is the number of reactant types. One can demonstrate this by induction from cases with few reactants. But for the cycle to exist materially, the constituents' rates of decay must equal their rate of creation from reagents. The rates of decay are increased by "OR-reactions" that tap the cycle [125], i.e. reactions where a constituent may undergo side-reactions. The factor limiting this creation rate is very likely to be one "limiting reagent". Exponential growth of the autocatalytic cycle will only occur when the limiting reagent (whichever one it is at the time) is present in excess, so for the cycle to persist, this limiting reagent must be generated in sufficient quantity. Reagents may be the products of other autocatalytic processes occurring under different equilibrium conditions in other micro-environments or may be due to non-autocatalytic recycling of some components of the system by solar radiation [126].

How specific must the reactions of an autocatalytic cycle be for it to grow? Imagine a cycle with n constituents, and m other active substances in the medium (which also include the reagents). Considering all possible reactions between the constituents and the other substances, King found that the cycle grows exponentially only if

$$\prod_{i=1}^{n} [1 + 1/S_i] \leq 2$$

for all n constituents, where S_i is the specificity of the ith constituent, given by

$$S_i = \alpha_i / \sum\nolimits_{j=1}^{m} \alpha_{ij}\,, j \neq i$$

where $\alpha_{ij} = \beta_{ij} R_j$, β_{ij} being the rate coefficient of the reaction between the ith constituent and the reactant j, and R_j being the concentration of the reactant j. In particular, the steady state of the limiting reagent R_L is given by

$$R_L = \alpha_L / \beta_L\,,$$

where β_L is the rate coefficient for uptake of the limiting reagent, and α_L is the rate of decay of the autocatalytic constituents.

How probable is it that a randomly generated autocatalytic cycle of size n will persist? Assuming randomly assigned rate coefficients and concentrations, King defines a "kinetic complexity" to a cycle $\big(Y = n(m-1)\big)$, where n is the number of constituents and m is the number of the active substances in the medium including reagents, and calculates the probability that a cycle of size n will persist under these conditions. King assumes an exponential distribution of specificities of reaction, with most reactions having low specificity. For an autocatalytic system with 4 uptake reactions and in a medium containing *only* the 4 appropriate reagents the chance of the cycle persisting is only 9×10^{-10}. See the table below for other values of Y.

Y	2	4	6	12	20	30
Probability	0.26	0.019	6×10^{-4}	9×10^{-10}	2×10^{-19}	2×10^{-33}

In conclusion, *selection of rate coefficients and concentrations of reagents are needed to make an autocatalytic cycle that persists with more than a very small number of constituents.* Random search will not do for anything but the smallest cycles.

Does an autocatalytic cycle conform to Maynard Smith's [1, 2] definition of a unit of evolution? Szathmáry classified autocatalytic cycles as replicators of the "holistic" type, and predicted that their heredity would be limited to a small number of alternative forms (basins of attraction in the chemical space of constituents), which showed only infrequent macromutations [22]. To what extent can autocatalytic cycles evolve as "holistic replicators" in chemical space? Obviously as a prerequisite, the cycle intermediates must not be lost, and therefore the limiting reagent must remain above its threshold at all times. King suggests selection would be largely confined to the specificity of the reaction for uptake of the limiting reagent. This could be achieved by loss of those materials that disrupted the recycling of the limiting reagent, or by exclusion of the m other species from the medium for which physical separation would be helpful. All else being equal, simpler autocatalytic cycles are easier to maintain. Separate autocatalytic cycles can compete for the same reactant, with competition in the growth phase being dependent on the rate of limiting reactant usage, and competition in the decay phase being dependent on the comparative decay rates [127]. Since growth is exponential, there is "survival of the fittest" during the growth phase, and co-existence is not possible, assuming a well-mixed reactor. During the decay phase there is exponential decay, with selection for autocatalytic particles with low decay rates. In a well-mixed reactor, co-existence can only occur if autocatalytic cycles are not competing for the same limiting reagent [127]. Co-operative interactions between autocatalytic cycles occur when their reactions are consecutive (i.e., the product of one is the reactant of the other) or where the constituent of one autocatalyst is the reagent for another autocatalyst. Such a coupling has been hypothesised by Kalapos, in which the formose cycle could have been anaplerotic of pyruvate (it supplied the limiting reagent) to the reductive citric acid cycle, therefore explaining the ubiquitous presence of the (methyl) glyoxalase pathway in living systems [128]. King has claimed that evolution from the first autocatalytic cycles to prokaryotes was due to a relatively small number of symbioses between cooperative autocatalytic cycles. Some sort of physical coupling between cycle constituents to form a combined "particle" would have been necessary in order for symbiosis to occur. King demonstrates that symbiosis would have been selectively advantageous when the limiting reagents of the original cycles were running low [126]. The chemoton is just such an example of three coupled autocatalytic cycles.

One crucial feature of the evolution of autocatalytic cycles is the bioenergetic constraint on the existence of continued recycling [122]. Evolution of metabolism seems to be rate-determined by its discovery of new "prime movers", just as cultural evolution seems to be rate-determined by the discovery of fire, water wheels, coal and nuclear power. An explanation of self-sustaining ecosystems of autocatalytic cycles must explain how novel energy sources were utilised for "complexification" of the metabolic network, using for example, electrical discharges, redox potentials, UV light, concentration by drying in intertidal zones, mineral surface films, or gradients across vesicles. Returning to the giant pre-biotic synthesis experiment, early attempts may have failed (i.e., have reached a point attractor of tar, or experienced the curse of combinatorial explosion [129]) because they used an ecosystem that was unable to exploit these novel energy sources: fatty acids aggregated and underwent the browning reaction, the mixture became disordered, pyrophosphates degraded, redox potentials were not utilisable, light energy could not be utilised by the constituents of the medium. So effectively *the ecosystem was not recycling*, it was not *collectively autotrophic* [130]. How can a sufficiently complex chemical network be recycling? Work has to be done on the network so that a constant flux of limiting reagents is available for autocatalytic cycles.

4.3
The Formose Cycle and the Reductive Citric Acid Cycle

Two candidates for the first pre-biological autocatalytic cycle are the formose cycle and the reductive citric acid cycle. The formose cycle was discovered by Butlerow [131] and has been investigated by others since [132–136]. A simplified version of the cycle is shown in Fig. 12. It converts formaldehyde to a mixture of complex molecules in alkaline (and only very slowly in neutral) solution containing a divalent metallic ion catalyst e.g. Ca^{2+} or Pb^{2+}. Two formaldehyde molecules form glycolaldehyde to enter the cycle but formaldehyde can also react by the Cannizzaro reaction in alkaline solution producing methanol and formic acid, so reducing the specificity of the critical reagent reaction. The reaction does not proceed exponentially until a limiting concentration of formaldehyde is exceeded, as expected from the models of King [127]. Glycolaldehyde is converted to glyceraldehyde which is then converted to tetrose, pentose and hexose sugars. The sugars decompose to hydroxy-acids and related compounds that are lost from the cycle, and may also interfere with the cycle. Many "mutant" related cycles exist that recycle various versions of the sugars. Each sugar can exist in D- or L-form, each enantiomer inhabiting its own version of the cycle. Important tapping side-reactions include the Cannizzaro reduction of sugars. The existence of these side-reactions increases the limiting concentration of formaldehyde required to run the cycle. If the kinetic conditions described previously are satisfied, two molecules of glycolaldehyde are produced for every one entering the cycle. If the concentration of formaldehyde is not maintained by continuous supply, the cycle runs down, but as long as it is maintained the formose cycle can be run in a flow reactor [132]. Leslie Orgel [136] describes the side-reactions of the formose reaction as "notorious". It is thought that formaldehyde occurred through atmospheric oxidation of methane [132]. How could this have occurred at high rates? How is it possible to reduce the side-reactions of the formose cycle, such that its constituents can persist at low formaldehyde concentration?

Wächtershäuser [11] argued that the reductive citric acid cycle (reductive tricorboxylaic acid cycle, rTCA) was the first autocatalytic cycle that evolved using the reducing power of pyrite surfaces. Smith and Morowitz [130] also argued that the rTCA cycle was a likely pre-biotic autocatalyst that could have worked without enzymes in the reducing atmosphere of the early earth. They propose it could have been a part of the "relaxation pathway" for the free energy "bottleneck" of redox couples created from volcanic magma. There are multiple possible synthetic pathways for sugars, lipids and amino acids, starting from the rTCA cycle, so possibly increasing their probability of evolution. Constituent reactions are first order so increasing turning rate, and acetate could be converted to lipids allowing vesicles to form [130]. These properties also apply to the formose cycle. They relate the energetic features of the rTCA cycle to its evolvability, "harmful side reactions that cannot be eliminated cost additional metabolic energy to handle. Thus, a metabolic core with high intrinsic efficiency and statistically favored reactions would in general leave more free-energy for the synthesis of higher-level regulatory structures than less intrinsically efficient alternatives."

4.4
Encapsulation of Autocatalytic Cycles

"Physical structure is an essential aspect of even the simplest autocatalysts in solution, ensuring their stability against side-reactions and enhancing their turnover rate. The kinetic characteristics of autocatalysis are such that, under some conditions, there may be feedback that alters the physical structure, building complexity [127]." For example, fire is autocatalytic and it spreads as a "front". Chemical waves are fronts of progressive energy dissipation. Bacterial growth is also autocatalytic, and spreads as a front. Spatial factors can have complex and non-linear effects on selection. These effects have been explored in models of co-evolution between metabolism and template replicators in silico [13] and for contemporary self-encoding macromolecular systems in vitro [51], but not for autocatalytic cycles, nor for autocatalytic cycles coupled with protocell growth under realistic conditions, e.g. side-reactions, and reagent re-cycling.

Encapsulation of metabolism in protocells could allow high metabolite concentrations to be maintained, especially those concentrations of large and charged molecules. It could allow increased specificity of reactions (by excluding poisons that can pass through the membrane), and selection of the autocatalytic cycle as a spatially discrete unit as well as a unit in chemical space. Limited membrane heredity could have operated with peptides and amphiphiles. On the road to vesicle encapsulation, intermediate steps might have been: encapsulation in less well-organised structures such as coacervates or diffusion limited on mineral surfaces. Ingeniously, Wächtershäuser's "bubble wrap" (abstriction) scenario involves the production of membrane caps by surface lipid metabolism of rTCA chemo-autotrophs, allowing the subsequent evolution of cytosol metabolism desorbed from the surface [11]. He proposes these could eventually produce vesicles. Wächtershäuser has argued that a heterotrophic origin in a prebiotic broth is implausible because cleavage reactions are favoured by increasing entropy; whereas on surfaces the increase in entropy caused by cleavage is small, so surface metabolism would be inherently synthetic.

The problem with both the formose cycle and the rTCA cycle is that we do not know how either could evolve to produce their own membrane constituents. Gánti had to hypothesise a separate reaction sequence from the formose cycle for synthesising the amphiphiles. Vesicles may be produced under primordial conditions, for example montmorillonite clay catalyses the formation of micelles which form into vesicles and trap some of the clay inside them [137]. However, demonstration of self-replicating vesicles capable of exponential growth has been limited to caprylic acid vesicles that catalyse the hydrolysis of ethyl caprylate into caprylic acid which then spontaneously incorporates into the vesicle causing growth [42]. Caprylic acid has no plausible pre-biological synthesis. It has not yet been demonstrated that an autocatalytic cycle can continuously produce amphiphiles that self-organise to enclose the metabolites at stoichiometric ratios that coordinate the exponential growth of the vesicle with that of the metabolism. What is the minimum metabolism, and the minimum synthetic procedure required to allow metabolism to enclose itself in a protocell that replicates exponentially and can evolve?

We can safely assume the existence of whole 'families' of vesicles without autocatalytic metabolisms that were not exponentially growing. Some of these vesicles may even have had H^+ gradients providing a source of energy for metabolism [17]. Some of these vesicles may have leant themselves to colonisation by autocatalytic metabolisms which had evolved to a significant degree already independently of the vesicles, e.g. on surfaces. However, in this case there would be the problem of impermeability. How would the highly evolved autocatalytic metabolism get into the vesicle? Although transient chemical transfer groups could solve this problem it seems unlikely these could occur all at once, and if they occurred incrementally, the autocatalytic cycle would presumably be compromised. Alternatively, it may be the case that evolution of autocatalytic metabolism may have been so difficult outside vesicles that even the earliest most inefficient autocatalytic metabolism required vesicles to evolve in; either vesicles produced by their own amphiphiles, or more likely, vesicles produced by other chemical systems. This may have been particularly beneficial if these vesicles possessed a method of utilising energy. Presumably a large amount of permeability variation would exist for selection to act upon. Heterogeneous membranes are pre-adapted for stability. At an early stage, perhaps walls of amphiphiles could separate compartments of autocatalytic cycles on a 2D surface, or vesicles could clump together in the presence of ions.

One experimental approach would be to generate heterogeneous vesicles in broths containing the constituents of the formose reaction, initially limiting the concentration of formaldehyde. After leaving the system for a while, formaldehyde would be added to the solution. Those vesicles that grew and divided fastest would be selected for. There may be rare conditions under which the constituents of the formose reaction are permeable to the membrane. They enter the vesicle and become trapped as soon as those conditions change. Formaldehyde could then enter the vesicles and allow the autocatalytic cycle to run.

Deamer [121] and others have advocated peptide evolution in early protocells capable of energy-transduction. He suggests the first membranes may have been made of monocarboxylic acids and alcohol. However, peptide evolution in protocells lacks any plausible mechanism for heredity of sequence. New sequences (e.g. coding for ligases or proteases) would have to be rediscovered in each lifetime.

4.5 Post-Enzymatic Evolution of Metabolism

Let us assume for now that an exponentially growing protocell with an enclosed autocatalytic metabolism could form and eventually evolve RNA enzymes. RNA enzymes would have co-evolved with the original metabolic pathways. After the evolution of protein enzymes, further takeover and transformation of pathways would have occurred. Pohorille and New [138] observed "since there is no relationship between the RNA catalytic power of a given RNA and the protein for which that RNA can code, there is no clear path from the RNA world to the protein world." Therefore, protein cladistics can only make conclusions about metabolism after protein enzymes have evolved [139]. To infer the state of living systems before protein enzymes, further theoretical assumptions are required.

The chemoton model suggests that a general motif of metabolism evolution was the stoichiometric coupling of reaction cycles and chains, shaped first by the underlying chemical bioenergetics, but shaped later by enzymes (leading to so-called catalytic supercycles [125]). The advantage of translated protein enzymes would have been the decoupling of the difficult process of their own self-replication from cross-catalysis. Vast catalytic spaces would become easily searchable by protein enzymes, after the underlying nucleic acid replication problem had been solved for all possible sequences. Unlimited heredity and microevolution of metabolism would be possible. Enzymes would only have been able to catalyse reactions for which at least one molecule of reactant existed. Only then could the enzyme have an effect. Therefore, *evolution of metabolism can be seen as a genetic assimilation of underlying metabolic pathways.* It is possible that non-functional enzymes (i.e. enzymes in cells without any suitable reactants) could exist, and "cultivate" the platonic space yet unexplored by the cells' metabolism.

Wächtershäuser [11] has produced a classification of elementary variational motifs describing evolution at the level of biochemical phenotypes (Scheme 1). Presumably because Wächtershäuser was interested in autotrophs, he did not consider the fundamental pathway operation of retro-evolution, discovered by Horowitz [140], which we include here as an additional fundamental operation **i**). Horowitz assumed that D, a complex organic molecule was present in the soup that heterotrophs first used, but that it ran out. Heterotrophs then evolved to use D's precursor C to synthesise D, and so on for B and A. Horowitz's mechanism is important if the end product D is indeed available at an early stage, and if C, B and A are available in excess in the environment also. This is only likely where autotrophs produce these compo-

a) **Terminal Extension:** $\rightarrow A \rightarrow B \Rightarrow \rightarrow A \rightarrow B \rightarrow C$

b) **Lateral Branching:** $\rightarrow A \rightarrow B \rightarrow C \Rightarrow \rightarrow A \rightarrow B \rightarrow C;\ B \rightarrow D$

c) Pathway Recruitment: $\rightarrow A \rightarrow B \rightarrow C \rightarrow D \Rightarrow A \rightarrow B \rightarrow C \rightarrow D;\ A \rightarrow X \Rightarrow A \rightarrow B \rightarrow C \rightarrow D;\ A \rightarrow X;\ X \rightarrow B' \rightarrow C' \rightarrow D'$

d) Pathway Abandonment: $A \rightarrow B \rightarrow C \Rightarrow A \rightarrow B \rightarrow C;\ A' \rightarrow B' \rightarrow C' \Rightarrow A' \rightarrow B' \rightarrow C'$

e) Pathway Takeover: $A \rightarrow B \Rightarrow A \rightarrow B;\ C \rightarrow B \Rightarrow C \rightarrow B$

f) Pathway Insertion: $A \rightarrow B \rightarrow C \Rightarrow A \rightarrow B \rightarrow C;\ D \rightarrow E \rightarrow F \rightarrow C; \Rightarrow A \rightarrow B \rightarrow C;\ A \rightarrow D \rightarrow E \rightarrow F \rightarrow G;$
$\Rightarrow A \rightarrow D \rightarrow E \rightarrow F \rightarrow C$

g) Retrograde Mimicry: $A \rightarrow B \rightarrow C \Rightarrow A \rightarrow B \rightarrow C;\ E \rightarrow D \rightarrow C; \Rightarrow E \rightarrow D \rightarrow C \Rightarrow E \rightarrow D \rightarrow C;\ F \rightarrow D \Rightarrow F \rightarrow D \rightarrow C$

h) Pathway Reversal: $A \rightarrow B \rightarrow C \Rightarrow C \rightarrow B \rightarrow A$.

Scheme 1 Classification of elementary variational motifs describing evolution at the level of biochemical phenotypes. Indicates enzyme catalysed chemical reactions, and indicates evolutionary progression. The fundamental operations are shown in bold nents, in which case retro-evolution may be an important mechanism where a heterotroph co-evolves closely with an autotroph.

i) **Retro-Extension:** B → C ⇒ A → B → C

It is assumed that loss of pathway components is also a fundamental operation **j)** that may occur due to stochastic loss during replication for example.

j) **Loss of Pathways:** A → B → C → D ⇒ A → B–ll

Using various combinations of these primitives, hypotheses can be formed regarding the evolutionary trajectory that led to extant metabolic systems. Because these primitives allow us to trace multiple pathways to extant metabolism, other assumptions must be made to limit the space of possible evolutionary trajectories. Melendez-Hevia et al. [141] suggest some additional assumptions of evolutionary "opportunism".

1. Any intermediate evolutionary metabolic pathway should be possible, albeit running slowly, without enzymes.
2. Intermediates should be stable to rapid decomposition, otherwise the pathway could not exist before enzymes or when only the earliest enzymes were present.
3. The existence of material inputs to the new pathway must be due to another metabolic process that had previously been selected for.
4. The new pathway cannot involve a reaction that is thermodynamically or kinetically incompatible with any other pathway in the same space simultaneously.

Using these constraints and the principle c) of pathway recruitment (Scheme 1), they argued that the Krebs cycle evolved opportunistically, originally as a biosynthetic pathway allowing synthesis of amino acids from pyruvate, and that this required the evolution of only one enzyme to catalyse the conversion of succinyl-CoA to succinate. This solution rests on the assumption that amino acid biosynthesis preceded the Krebs cycle, for which there is evidence [139]. They also suggested several alternative mechanisms for catabolising acetate groups (the current role of the Krebs cycle) and showed that these would have been more difficult to evolve or less efficient, given the principles of opportunism.

The question whether there is a historical trace of the retroevolution of pathways [140] or another alternative, the so-called patchwork mechanism [142] has been asked repeatedly. The first evolution scenario predicts that enzymes catalysing steps of the same metabolic pathway should be phylogenetically related (homologous). The latter scenario states that enzymes evolve by increasing their specificity: first enzyme E catalyses reactions of chemically similar substrates S1 and S2, then its genes duplicates and the two enzymes diverge, E1 and E2 evolving specificity towards S1 and S2, respectively. Thus homologous enzymes should be functionally similar, but they may or may not catalyse neighbouring reactions in metabolic pathways. Skipping the whole history of more and more refined analyses, we just mention the latest result.

Light and Kraulis [143] found, analysing the complete available dataset from *E. coli*, that homologous enzyme pairs abound at the minimal path length of one (i.e. the product of one is the substrate of the other, or vice versa). This may corroborate the retroevolution scenario. Not so, for two reasons. First, there is a small degree of homology between enzyme pairs with a mean path length of 2, and negligible homology between a pair of mean path length of 3 or greater. Second, the majority of homologous pairs with mean path length of 1 have similar EC numbers, hence they are functionally related. Therefore, the most recent analysis seems to support the patchwork model of enzyme evolution.

Note that analysis of contemporary protein enzymes do not necessarily shed light on the primordial build-up of metabolic networks, if modern metabolism is a palimpsest of the RNA world [144]. If there was a metabolically rich RNA world, then a large part of the network must have been built up before the advent of encoded proteins. By the time of the origin of translation primary heterotrophy (if there was one), feeding on the prebiotic environment, must have been over. Hence the fact that we find no strong evidence for retroevolution in contemporary enzymatic metabolism may say nothing about its original significance in the RNA world.

Let us point out a perhaps even more severe objection. Historical retroevolution is fully compatible with the patchwork type of enzyme recruitment, provided at least two pathways are present and evolving in parallel. The higher the number of pathways, the more "patchworky" the recruitment can be, even though each individual pathway is retroevolving.

It is also important to realise that the Horowitz scenario makes sense only if, although intermediates of contemporary pathways have been present without enzymatic aid in the milieu of protocells, they were synthesised elsewhere. In contrast, if they could be chemically synthesised in the same milieu, then autotrophy would have been easy simply by letting the reactions run *inside* protocells. It is no miracle that in Wächtersäuser's [10, 11] scenario, where everything is formed in situ on the pyrite surface, autotrophy is given for free.

Following the considerable recent interest in scale-free networks, Jeong et al. [145] have shown that the metabolic networks of extant living systems are "scale-free networks" sharing the same metabolite "hubs" over evolutionary time. Wagner and Fell [146] suggest that there are three reasons why metabolic networks may be scale free.

1. Metabolism may be scale-free because of chemical constraints of the underlying chemical network. They dismiss this possibility, claiming as evidence the fact that in different organisms the metabolic network takes many different forms. However, this does not rule out the existence of more general chemical constraints that may produce the scale-free network. For example, it is generally the case that small molecules have more possible synthesis routes than large molecules and so we expect connectivity to scale as a function of size.

2. Metabolism may be scale free because during evolution, metabolites with more connections are more likely to make further connections, implying that older metabolites are those with the largest number of connections. This is based on Barabási's algorithm for producing scale-free networks [147]. Morowitz uses the assumption that older metabolites should be more highly connected to support the idea of an early amino acid metabolism. The metabolites with highest connectivity are glutamate, pyruvate, and coenzyme A. However, this simple algorithm for constructing scale-free networks is one of an infinite number of methods, many of which involve pruning as well as addition of connections or metabolites, e.g. it is likely that pathway recruitment c) (Scheme 1) would tend to produce networks with "scale-free" properties. This would mean that it is not possible to conclusively infer phylogenetic age from connectivity.
3. Metabolism may be scale-free as a result of selection for functional properties, i.e. the rapid propagation of perturbations resulting in developmental robustness and evolvability. Jeong et al. [145] support this adaptive interpretation. They showed that the diameter of the network, i.e. the average path distance from one metabolite to another, did not increase with the number of metabolites in a given organism. They claimed that scale-free metabolic networks should be more evolvable than random metabolic networks because deleting nodes randomly (i.e., randomly removing metabolites) would not have a great effect on network diameter, and that networks with larger diameters would reduce the organisms' ability to respond effectively to external changes or internal errors because "offsetting these changes would involve a longer alternative biochemical pathway and consequently the synthesis of more new enzymes than within a metabolic network with a smaller diameter." However, there is no evidence relating network diameter and rate of reaction to perturbations, to evolvability. Bioinformatic evidence is against this explanation. Metabolic network analysis of yeast has provided fascinating results [148]. First, the majority of genes that looked dispensable turn out to be such only under laboratory conditions. Second, gene duplicates catalysing the same reaction are not more common for indispensable reactions, suggesting that the reason for their retention is not to provide compensation; instead, their presence is better explained by selection for high enzymatic flux. Third, only 4–17% of *in silico* deleted, dispensable gene products are buffered by metabolic network flux reorganisation. In fact a different, more chemically minded adaptive reasoning may turn out to be more fruitful. The main hubs in metabolism are molecules like water and the coenzymes. Coenzymes carry important functional groups, thus connecting different pathways in the network. It seems rational to argue that this is a better adaptive design than running a network without them. Coenzymes (like ATP) play the same role in metabolism as money does in economy.

More work is required to understand the adaptive significance of the small-world character in experiments in which the above explanations can be independently controlled in an evolutionary model of dynamic autocatalytic metabolism.

5 Outlook

We have witnessed spectacular development revolving around the minimal life/protocell idea in the last few years, although the roots of this vision go back to more than thirty years ago. A fairly unusual interplay between theory and experiment seems to unfold: theoretical models investigate possible dynamics, and experiments either confirm them or come up with surprising, unexpected novelties, but increasingly often reflecting back on theoretical considerations.

Most experimental studies attempt at the synthesis of infrabiological systems, i.e. all three subsystems (metabolism, template polycondensation, and membrane) of the chemoton are not yet figuring in those trials: practically everybody is now concentrating on membrane/template systems. Metabolism seems to be an especially hard problem, since we do not know where a sufficient amount of channelling could have come from without enzymes. We strongly urge the initiation of another experimental line, aiming at a metabolism/membrane infrabiological system, preferably using the formose reaction for metabolism.

One cannot deny the fact that the origin of unlimited heredity is an unsolved problem. Perhaps compartmentation will help solve this problem as well, so that long templates could self-replicate within vesicles without enzymatic aid.

We emphasise that top-down approaches to a "minimal genome" do not solve our problem: the spontaneous generation of cells with several hundred genes can be safely ruled out. We must adopt a bottom-up strategy instead: this is exactly the attempted synthesis of various infrabiological systems. The ultimate goal is, of course, to arrive at a chemical supersystem which at the same time is a *bona fide* biological system, conforming to Gánti's chemoton model.

Another top-down approach, namely phylogenetic analysis of contemporary protein enzymes is also of limited help. Very early primordial systems must have been enzyme-free, and later early systems could have been catalysed by ribozymes. Consequently, a substantial part of basic metabolism could have originated in an era about which there simply cannot be memories in protein coding genes.

Synthesis of a living chemical system may not shed too much light on the historical process of the origination of life, but we are optimistic that work in progress will contribute to the solution of one of the outstanding unsolved problems of science.

Acknowledgements Comments by Kepa Ruiz Mirazo have been helpful in revising a previous draft of the manuscript. E.Sz. is supported by the Hungarian Scientific Research Fund (OTKA T047295). M.S. is supported by Fundación Ramón Areces (Spain). The sponsorship from COST-European Cooperation in the field of Scientific and Technical Research, Internal Project number D27/0003/02 "Emergence and Selection of Networks of Catalytic Species" (URL:http://cost.cordis.lu/src/extranet/publish/D27WGP/d27-0003-02.cfm), is very much appreciated.

References

1. Maynard Smith J (1986) The problems of biology. Oxford University Press, Oxford, UK
2. Maynard Smith J (1987) How to model evolution. In: Dupré J (ed) The latest on the best: essays on evolution and optimality. MIT Press, Cambridge, MA, pp 119–31
3. Dawkins R (1976) The selfish gene. Oxford University Press, Oxford, UK
4. Gánti T (1971) Az élet princípuma (The principle of life), 1st ed. Gondolat, Budapest, Hungary
5. Gánti T (2003) The principles of life. Oxford University Press, Oxford, UK
6. Szathmáry E (2002) Units of evolution and units of life. In: Pályi G, Zucchi L, Caglioti L (eds) Fundamentals of Life. Elsevier, Paris, pp 181–195
7. Maynard Smith J, Szathmáry E (1995) The major transitions in evolution. Oxford University Press, Oxford, UK
8. Gilbert W (1986) Nature 319:618
9. Bernal JD (1967) The origin of life. Weidenfield, Nicolson, London
10. Wächtershäuser G (1988) Microbiol Rev 52:452
11. Wächtershäuser G (1992) Prog Biophys Mol Biol 58:85
12. Franchi M, Gallori E (2004) Orig Life Evol Biosph 34:133
13. Scheuring I, Czárán T, Szabó P, Károlyi G, Toroczkai Z (2003) Orig Life Evol Biosph 33:319
14. Wade MJ (1978) Quart Rev Biol 53:101
15. Boorman SA, Levitt PR (1980) The genetics of altruism. Academic Press, New York
16. Gánti T (1978) Az élet princípuma (The principle of life), 2nd (revised) ed. Gondolat, Budapest, Hungary
17. Szostak JW, Bartel DP, Luisi PL (2001) Nature 409:387
18. Pohorille A, Deamer D (2002) Trends Biotechnol 20:123
19. Szathmáry E (2003) The biological significance of Gánti's work in 1971 and today. In: Gánti T (ed) The principles of life. Oxford University Press, Oxford, UK, pp 157–168
20. Gánti T (1974) Biológia 22:17
21. Cavalier-Smith T (2000) Trends Plant Sci 5:174
22. Szathmáry E (2000) Phil Trans R Soc Lond B 355:1669
23. Segré D, Ben-Eli D, Deamer DW, Lancet D (2001) Orig Life Evol Biosph 31:119
24. Segré D, Ben-Eli D, Lancet D (2000) Proc Natl Acad Sci USA 97:4112
25. Kauffman SA (1986) J Theor Biol 119:1

26. Lee DH, Severin K, Yokobayashi Y, Ghadiri MR (1997) Nature 390:591. Erratum (1998) Nature 394:101
27. Fendler HJ, Fendler EJ (1975) Catalysis in micellar and macromolecular systems. Academic Press, New York
28. Lancet D, Sadovsky E, Seidemann E (1993) Proc Natl Acad Sci USA 90:3715
29. Hogeweg P (1998) On searching generic properties of non generic phenomena: an approach to bioinformatic theory formation. In: Adami C, Belew RK, Kitano H, Taylor CE (eds) Artificial life VI. MIT Press, Cambridge MA, pp 285–294
30. Segré D, Shenhav B, Kafri R, Lancet D (2001) J Theor Biol 213:481
31. Luisi PL, Stano P, Rasi S, Mavelli F (2004) Artif Life 10:297
32. Rashevsky N (1938) Mathematical biophysics. University of Chicago Press, Chicago
33. Gánti T, Gáspár C (1978) Acta Chim Acad Sci Hung 98:278
34. Verhás J (1989) Appendix. In: Gánti T (ed) Chemoton elmélet II (Chemoton theory Vol 2). OMIKK, Budapest, Hungary, pp 201–216
35. Tarumi K, Schwegler H (1987) Bull Math Biol 49:307
36. Božič B, Svetina S (2004) Eur Biophys J 33:565
37. Kaler WK, Murthy AK, Rodríguez BE, Zasadzinski JAN (1989) Science 245:1371
38. Lipowsky R (1991) Nature 349:475
39. Rokhsar DS, Anderson PW, Stein DL (1986) J Mol Evol 23:119
40. Döbereiner H-G, Käs J, Noppl D, Sprenger I, Sackmann E (1993) Biophys J 65:1396
41. Luisi PL, Walde P, Oberholzer T (1999) Curr Op Coll Interf Sci 4:33
42. Bachmann PA, Luisi PL, Lang J (1992) Nature 357:57
43. Blöchliger E, Blocher M, Walde P, Luisi PL (1998) J Phys Chem 102:10383
44. Rasi PS, Mavelli F, Luisi PL (2004) Orig Life Evol Biosph 34:243
45. Berclaz N, Müller M, Walde P, Luisi PL (2001) J Phys Chem B 105:1056
46. Berclaz N, Blöchliger E, Müller M, Luisi PL (2001) Phys J Chem B 105:1065
47. Walde P, Goto A, Monnard PA, Wessicken, Luisi PL (1994) J Am Chem Soc 116:7541
48. Walde P, Wick R, Fresta M, Mangone A, Luisi PL (1994) J Am Chem Soc 116:11649
49. Oberholzer T, Wick R, Luisi PL, Biebricher CK (1995) Biochem Biophys Res Commun 207:250
50. Oberholzer T, Albrizio M, Luisi PL (1995) Chem Biol 2:677
51. Hitz T, Luisi PL (2000) Biopolymers 55:381
52. Mavelli F, Luisi PL (1996) J Phys Chem 100:16600
53. Hanczyc MM, Fujikawa SM, Szostak JW (2003) Science 302:618
54. Chen IA, Roberts RW, Szostak JW (2004) Science 305:1474
55. Rasmussen S, Chen L, Nilsson M, Abe S (2003) Artif Life 9:269
56. Rasmussen SJ, Chen L, Stadler BM, Stadler PF (2004) Orig Life Evol Biosph 34:171
57. Nielsen PE (1993) Orig Life Evol Biosph 23:323
58. Szathmáry E, Gladkih I (1989) J Theor Biol 138:55
59. Stillwell W (1976) Biosystems 8:111
60. Stillwell W, Rau A (1981) Orig Life 11:243
61. Chakrabarti AC, Deamer DW (1994) J Mol Evol 39:1
62. Chen IA, Szostak JW (2004) Proc Natl Acad Sci USA 101:7965
63. Eigen M (1971) Naturwissenschaften 58:465
64. Friedberg EC, Walker GC, Siede W (1995) DNA repair and mutagenesis. ASM Press, Washington
65. Johnston WK, Unrau PJ, Lawrence MS, Glasner ME, Bartel DP (2001) Science 292:1319
66. Eigen M, Schuster P (1979) The hypercycle: a principle of natural self-organization. Springer, Berlin Heidelberg New York

67. Eigen M, Schuster P (1982) J Mol Evol 19:47
68. Nowak MA (1992) Trends Ecol Evol 7:118
69. Alves D, Fontanari JF (1998) Phys Rev E 57:7008
70. Eigen M, Schuster P (1978) Naturwissenschaften 65:341
71. Zintzaras E, Santos M, Szathmáry E (2002) J Theor Biol 217:167
72. Nuño JC, Tarazona P (1994) Bull Math Biol 56:875
73. Boerlijst MC, Hogeweg P (1991) Physica D 48:17
74. Maynard Smith J (1979) Nature 280:445
75. Eigen M, Schuster P, Gardiner W, Winkler-Oswatitsch R (1981) Sci Am 244:78
76. Bresch C, Niesert U, Harnasch D (1980) J Theor Biol 85:399
77. Luisi PL (1998) Orig Life Evol Biosph 28:613
78. Matsuura T, Yamaguchi M, Ko-Mitamura EP, Shima Y, Urabe I (2002) Proc Nat Acad Sci USA 99:7514
79. Szathmáry E (2003) Orig Life Evol Biosph 33:313
80. Niesert U, Harnasch D, Bresch C (1981) J Mol Evol 17:348
81. Szathmáry E, Demeter L (1987) J Theor Biol 128:463
82. Niesert U (1987) Orig Life 17:155
83. Suzuki H, Ono N (2003) Proc First Australian Conf Artif Life (ACAL-2003)
84. Santos M, Zintzaras E, Szathmáry E (2003) Orig Life Evol Biosph 33:405
85. Fontanari JF, Santos M, Szathmáry E (2005) J Theor Biol (in press)
86. Hogeweg P, Takeuchi N (2003) Orig Life Evol Biosph 33:375
87. Lehman N (2003) J Mol Evol 56:770
88. Chetverina HV, Demidenko AA, Ugarov VI, Chetverin (1999) FEBS Lett 450:89
89. Riley CA, Lehman N (2003) Chem Biol 10:1233
90. Santos M, Zintzaras E, Szathmáry E (2004) J Mol Evol 53:507
91. Temin HM (1993) Proc Natl Acad Sci USA 90:6900
92. Negroni M, Buc H (2000) Proc Natl Acad Sci USA 97:6385
93. Hwang CK, Svarovskaia ES, Pathak VK (2001) Proc Natl Acad Sci USA 98:12209
94. Moumen A, Polomack L, Roques B, Buc H, Negroni M (2001) Nucl Acids Res 29:3814
95. Cheng CP, Nagy PD (2003) J Virol 77:12033
96. Cavalier-Smith T (2002) Heredity 88:125
97. Maynard Smith J, Szathmáry E (1993) J Theor Biol 164:437
98. Koch AL (1984) J Mol Evol 20:71
99. Ohno S (1979) Evolution by gene duplication. Springer, Berlin Heidelberg New York
100. Prescott DM (1994) Microbiol Rev 58:233
101. Summers D (1998) Mol Microbiol 29:1137
102. Zhang Z, Cavallier-Smith T, Green BR (2002) Mol Biol Evol 19:489
103. Hackett JD, Yoon HS, Soares MB, Bonaldo MF, Casavant TL, Scheetz TE, Nosenko T, Bhattacharya D (2004) Curr Biol 14:213
104. Baird SE, Klobutcher LA (1991) J Protozool 38:136
105. Skovorodkin IN, Zassoukhina IB, Hojak S, Ammermann D, Gunzl A (2001) Chromosoma 110:352
106. Weiner AM, Maizels N (1987) Proc Natl Acad Sci USA 84:383
107. Paulsson J (2002) Genetics 161:1373
108. Michod RE, Levin BR (1988) The evolution of sex: an examination of current ideas. Sinauer, Sunderland, MA
109. Hamilton WD, Axelrod R, Tanese R (1990) Proc Natl Acad Sci USA 87:3566
110. Woese C (1998) Proc Natl Acad Sci USA 95:6854
111. Rivera MC, Lake JA (2004) Nature 431:152
112. Koonin EV (2000) Annu Rev Genomics Hum Genet 1:99

113. Gil R, Silva FJ, Peretó J, Moya A (2004) Microb Mol Biol Rev 68:518
114. Oparin AI (1924) Proiskhozdenie zhizni. Izd. Moskovshii Rabochii, Moscow. English translation by Synge A, printed In: Bernal JD (ed) (1967) The origin of life. Weidenfeld, Nicholson London. World Publishing Co Cleveland, New York, pp 199–234
115. Haldane JBS (1929) The origin of life. The Rationalist Annual. Reprinted in: Bernal JD (ed) (1967) The origin of life. World Publishing Co, Cleveland, New York, pp 242–249
116. Farmer JD, Kauffman SA, Packard NH (1986) Physica 22D:50
117. Fox SW (1965) Nature 205:336
118. Kleidon A (2004) Climatic Change 66:271
119. Dyson FJ (1999) Origins of life. Cambridge Univ Press, Cambridge
120. Prigogine I (1977) Self-organization in nonequilibrium systems: from dissipative structures to order through fluctuations. Wiley, New York
121. Deamer DW (1987) Microbiol Mol Biol Rev 61:239
122. Morowitz H (1977) Adv Biol Med Phys 16:151
123. Saier MH (2000) J Bacter 182:5029
124. King GAM (1982) Biosystems 15:89
125. Ganti T (2004) Chemoton theory, Vol I & Vol II. Kluwer, Dordrecht, The Netherlands
126. King GAM (1977) Orig Life 8:38
127. King GAM (1978) Chem Soc Rev 7:297
128. Kalapos MP (1999) Exp Mol Med 31:1
129. Schuster P (2000) Proc Natl Acad Sci USA 97:7678
130. Smith E, Morowitz HJ (2004) Proc Natl Acad Sci USA 101:13 168
131. Butlerow A (1861) Ann Chem 120:295
132. Decker P, Schweer H, Pohlmann R (1982) J Chrom 244:281
133. Mizano T, Weiss AH (1974) Adv Carbohyd Chem Biochem 29:173
134. Socha R, Weiss AH, Sakharow MM (1980) React Kinet Catal Lett 14:119
135. Breslow R (1959) Tetrahedron Lett 21:22
136. Orgel L (2004) Crit Rev Bioch Molr Biol 39:99
137. Hanczyc MM, Fujikawa SM, Szostak JW (2003) Science 302:618
138. Pohorille A, New N (2002) Models of protocellular structures, functions and evolution (Abstract). In: Proc of the XIIth Rencontres de Blois: Frontiers of Life Conference, 2000, Chateau de Blois, France
139. Cunchillos C, Lecointre G (2003) J Biol Compl 278:47 960
140. Horowitz NH (1945) Proc Natl Acad Sci USA 31:153
141. Melendez-Hevia E, Waddell TG, Cascante M (1996) J Mol Evol 43:293
142. Jensen RA (1976) Annu Rev Microbiol 30:409
143. Light S, Kraulis P (2004) BMC Bioinformatics 5:15 (doi:10.1186/1471-2105-5-15)
144. Benner SA, Ellington AD, Tauer A (1989) Proc Natl Acad Sci USA 86:7054
145. Jeong H, Tombor B, Albert R, Oltval ZN, Barabasi AL (2000) Nature 407:651
146. Wagner A, Fell DA (2000) Santa Fe Institute Tech Rep 00-07-041
147. Barabási AL (1999) Los Alamos Nat Lab Prep cond-mat/9907068
148. Papp B, Pal C, Hurst LD (2004) Nature 429:661
149. Ruiz-Mirazo, Moreno A (2004) Artif Life 10:235
150. Von Kiedrowski G (1986) Angew Chem Int Ed 25:932
151. Von Kiedrowski G (1993) Bioorg Chem Front 3:113

Author Index Volumes 251–259

Author Index Vols. 26–50 see Vol. 50
Author Index Vols. 51–100 see Vol. 100
Author Index Vols. 101–150 see Vol. 150
Author Index Vols. 151–200 see Vol. 200
Author Index Vols. 201–250 see Vol. 250

The volume numbers are printed in italics

Ajayaghosh A, George SJ, Schenning APHJ (2005) Hydrogen-Bonded Assemblies of Dyes and Extended π-Conjugated Systems. *258*: 83–118
Alberto R (2005) New Organometallic Technetium Complexes for Radiopharmaceutical Imaging. *252*: 1–44
Anderson CJ, see Li WP (2005) *252*: 179–192
Anslyn EV, see Houk RJT (2005) *255*: 199–229
Araki K, Yoshikawa I (2005) Nucleobase-Containing Gelators. *256*: 133–165
Armitage BA (2005) Cyanine Dye–DNA Interactions: Intercalation, Groove Binding and Aggregation. *253*: 55–76
Arya DP (2005) Aminoglycoside–Nucleic Acid Interactions: The Case for Neomycin. *253*: 149–178

Bailly C, see Dias N (2005) *253*: 89–108
Balaban TS, Tamiaki H, Holzwarth AR (2005) Chlorins Programmed for Self-Assembly. *258*: 1–38
Barbieri CM, see Pilch DS (2005) *253*: 179–204
Bayly SR, see Beer PD (2005) *255*: 125–162
Beer PD, Bayly SR (2005) Anion Sensing by Metal-Based Receptors. *255*: 125–162
Boiteau L, see Pascal R (2005) *259*: 105–176
Boschi A, Duatti A, Uccelli L (2005) Development of Technetium-99m and Rhenium-188 Radiopharmaceuticals Containing a Terminal Metal–Nitrido Multiple Bond for Diagnosis and Therapy. *252*: 85–115
Braga D, D'Addario D, Giaffreda SL, Maini L, Polito M, Grepioni F (2005) Intra-Solid and Inter-Solid Reactions of Molecular Crystals: a Green Route to Crystal Engineering. *254*: 71–94
Brizard A, Oda R, Huc I (2005) Chirality Effects in Self-assembled Fibrillar Networks. *256*: 167–218

Chaires JB (2005) Structural Selectivity of Drug-Nucleic Acid Interactions Probed by Competition Dialysis. *253*: 33–53
Chiorboli C, Indelli MT, Scandola F (2005) Photoinduced Electron/Energy Transfer Across Molecular Bridges in Binuclear Metal Complexes. *257*: 63–102
Collyer SD, see Davis F (2005) *255*: 97–124
Commeyras A, see Pascal R (2005) *259*: 105–176
Correia JDG, see Santos I (2005) *252*: 45–84
Costanzo G, see Saladino R (2005) *259*: 29–104
Crestini C, see Saladino R (2005) *259*: 29–104

D'Addario D, see Braga D (2005) *254*: 71–94
Davis F, Collyer SD, Higson SPJ (2005) The Construction and Operation of Anion Sensors: Current Status and Future Perspectives. *255*: 97–124
Deamer DW, Dworkin JP (2005) Chemistry and Physics of Primitive Membranes. *259*: 1–27
Dervan PB, Poulin-Kerstien AT, Fechter EJ, Edelson BS (2005) Regulation of Gene Expression by Synthetic DNA-Binding Ligands. *253*: 1–31
Dias N, Vezin H, Lansiaux A, Bailly C (2005) Topoisomerase Inhibitors of Marine Origin and Their Potential Use as Anticancer Agents. *253*: 89–108
DiMauro E, see Saladino R (2005) *259*: 29–104
Dobrawa R, see You C-C (2005) *258*: 39–82
Duatti A, see Boschi A (2005) *252*: 85–115
Dworkin JP, see Deamer DW (2005) *259*: 1–27

Edelson BS, see Dervan PB (2005) *253*: 1–31
Edwards DS, see Liu S (2005) *252*: 193–216
Escudé C, Sun J-S (2005) DNA Major Groove Binders: Triple Helix-Forming Oligonucleotides, Triple Helix-Specific DNA Ligands and Cleaving Agents. *253*: 109–148

Fages F, Vögtle F, Žinić M (2005) Systematic Design of Amide- and Urea-Type Gelators with Tailored Properties. *256*: 77–131
Fages F, see Žinić M (2005) *256*: 39–76
Fechter EJ, see Dervan PB (2005) *253*: 1–31
Fernando C, see Szathmáry E (2005) *259*: 255–299
De Feyter S, De Schryver F (2005) Two-Dimensional Dye Assemblies on Surfaces Studied by Scanning Tunneling Microscopy. *258*: 205–255
Fujiwara S-i, Kambe N (2005) Thio-, Seleno-, and Telluro-Carboxylic Acid Esters. *251*: 87–140

Gelinck GH, see Grozema FC (2005) *257*: 135–164
George SJ, see Ajayaghosh A (2005) *258*: 83–118
Giaffreda SL, see Braga D (2005) *254*: 71–94
Grepioni F, see Braga D (2005) *254*: 71–94
Grozema FC, Siebbeles LDA, Gelinck GH, Warman JM (2005) The Opto-Electronic Properties of Isolated Phenylenevinylene Molecular Wires. *257*: 135–164

Higson SPJ, see Davis F (2005) *255*: 97–124
Hirst AR, Smith DK (2005) Dendritic Gelators. *256*: 237–273
Holzwarth AR, see Balaban TS (2005) *258*: 1–38
Houk RJT, Tobey SL, Anslyn EV (2005) Abiotic Guanidinium Receptors for Anion Molecular Recognition and Sensing. *255*: 199–229
Huc I, see Brizard A (2005) *256*: 167–218

Ihmels H, Otto D (2005) Intercalation of Organic Dye Molecules into Double-Stranded DNA – General Principles and Recent Developments. *258*: 161–204
Indelli MT, see Chiorboli C (2005) *257*: 63–102
Ishii A, Nakayama J (2005) Carbodithioic Acid Esters. *251*: 181–225
Ishii A, Nakayama J (2005) Carboselenothioic and Carbodiselenoic Acid Derivatives and Related Compounds. *251*: 227–246
Ishi-i T, Shinkai S (2005) Dye-Based Organogels: Stimuli-Responsive Soft Materials Based on One-Dimensional Self-Assembling Aromatic Dyes. *258*: 119–160

James DK, Tour JM (2005) Molecular Wires. *257*: 33–62
Jones W, see Trask AV (2005) *254*: 41–70

Kambe N, see Fujiwara S-i (2005) *251*: 87–140
Kano N, Kawashima T (2005) Dithiocarboxylic Acid Salts of Group 1–17 Elements (Except for Carbon). *251*: 141–180
Kato S, Niyomura O (2005) Group 1–17 Element (Except Carbon) Derivatives of Thio-, Seleno- and Telluro-Carboxylic Acids. *251*: 19–85
Kato S, see Niyomura O (2005) *251*: 1–12
Kato T, Mizoshita N, Moriyama M, Kitamura T (2005) Gelation of Liquid Crystals with Self-Assembled Fibers. *256*: 219–236
Kaul M, see Pilch DS (2005) *253*: 179–204
Kaupp G (2005) Organic Solid-State Reactions with 100% Yield. *254*: 95–183
Kawashima T, see Kano N (2005) *251*: 141–180
Kitamura T, see Kato T (2005) *256*: 219–236
Komatsu K (2005) The Mechanochemical Solid-State Reaction of Fullerenes. *254*: 185–206
Kriegisch V, Lambert C (2005) Self-Assembled Monolayers of Chromophores on Gold Surfaces. *258*: 257–313

Lahav M, see Weissbuch I (2005) *259*: 177–254
Lambert C, see Kriegisch V (2005) *258*: 257–313
Lansiaux A, see Dias N (2005) *253*: 89–108
Leiserowitz L, see Weissbuch I (2005) *259*: 177–254
Lhoták P (2005) Anion Receptors Based on Calixarenes. *255*: 65–95
Li WP, Meyer LA, Anderson CJ (2005) Radiopharmaceuticals for Positron Emission Tomography Imaging of Somatostatin Receptor Positive Tumors. *252*: 179–192
Liu S (2005) 6-Hydrazinonicotinamide Derivatives as Bifunctional Coupling Agents for ^{99m}Tc-Labeling of Small Biomolecules. *252*: 117–153
Liu S, Robinson SP, Edwards DS (2005) Radiolabeled Integrin $\alpha_v\beta_3$ Antagonists as Radiopharmaceuticals for Tumor Radiotherapy. *252*: 193–216
Liu XY (2005) Gelation with Small Molecules: from Formation Mechanism to Nanostructure Architecture. *256*: 1–37

Maini L, see Braga D (2005) *254*: 71–94
Matsumoto A (2005) Reactions of 1,3-Diene Compounds in the Crystalline State. *254*: 263–305
Meyer LA, see Li WP (2005) *252*: 179–192
Mizoshita N, see Kato T (2005) *256*: 219–236
Moriyama M, see Kato T (2005) *256*: 219–236
Murai T (2005) Thio-, Seleno-, Telluro-Amides. *251*: 247–272

Nakayama J, see Ishii A (2005) *251*: 181–225
Nakayama J, see Ishii A (2005) *251*: 227–246
Niyomura O, Kato S (2005) Chalcogenocarboxylic Acids. *251*: 1–12
Niyomura O, see Kato S (2005) *251*: 19–85

Oda R, see Brizard A (2005) *256*: 167–218
Otto D, see Ihmels H (2005) *258*: 161–204

Pascal R, Boiteau L, Commeyras A (2005) From the Prebiotic Synthesis of α-Amino Acids Towards a Primitive Translation Apparatus for the Synthesis of Peptides. *259*: 105–176
Paulo A, see Santos I (2005) *252*: 45–84
Pilch DS, Kaul M, Barbieri CM (2005) Ribosomal RNA Recognition by Aminoglycoside Antibiotics. *253*: 179–204
Piwnica-Worms D, see Sharma V (2005) *252*: 155–178
Polito M, see Braga D (2005) *254*: 71–94
Poulin-Kerstien AT, see Dervan PB (2005) *253*: 1–31

Ratner MA, see Weiss EA (2005) *257*: 103–133
Robinson SP, see Liu S (2005) *252*: 193–216

Saha-Möller CR, see You C-C (2005) *258*: 39–82
Sakamoto M (2005) Photochemical Aspects of Thiocarbonyl Compounds in the Solid-State. *254*: 207–232
Saladino R, Crestini C, Costanzo G, DiMauro E (2005) On the Prebiotic Synthesis of Nucleobases, Nucleotides, Oligonucleotides, Pre-RNA and Pre-DNA Molecules. *259*: 29–104
Santos I, Paulo A, Correia JDG (2005) Rhenium and Technetium Complexes Anchored by Phosphines and Scorpionates for Radiopharmaceutical Applications. *252*: 45–84
Santos M, see Szathmáry E (2005) *259*: 255–299
Scandola F, see Chiorboli C (2005) *257*: 63–102
Scheffer JR, Xia W (2005) Asymmetric Induction in Organic Photochemistry via the Solid-State Ionic Chiral Auxiliary Approach. *254*: 233–262
Schenning APHJ, see Ajayaghosh A (2005) *258*: 83–118
Schmidtchen FP (2005) Artificial Host Molecules for the Sensing of Anions. *255*: 1–29 Author Index Volumes 251–255
De Schryver F, see De Feyter S (2005) *258*: 205–255
Sharma V, Piwnica-Worms D (2005) Monitoring Multidrug Resistance P-Glycoprotein Drug Transport Activity with Single-Photon-Emission Computed Tomography and Positron Emission Tomography Radiopharmaceuticals. *252*: 155–178
Shinkai S, see Ishi-i T (2005) *258*: 119–160
Siebbeles LDA, see Grozema FC (2005) *257*: 135–164
Smith DK, see Hirst AR (2005) *256*: 237–273
Stibor I, Zlatušková P (2005) Chiral Recognition of Anions. *255*: 31–63
Suksai C, Tuntulani T (2005) Chromogenetic Anion Sensors. *255*: 163–198
Sun J-S, see Escudé C (2005) *253*: 109–148
Szathmáry E, Santos M, Fernando C (2005) Evolutionary Potential and Requirements for Minimal Protocells. *259*: 255–299

Tamiaki H, see Balaban TS (2005) *258*: 1–38
Tobey SL, see Houk RJT (2005) *255*: 199–229
Toda F (2005) Thermal and Photochemical Reactions in the Solid-State. *254*: 1–40
Tour JM, see James DK (2005) *257*: 33–62
Trask AV, Jones W (2005) Crystal Engineering of Organic Cocrystals by the Solid-State Grinding Approach. *254*: 41–70
Tuntulani T, see Suksai C (2005) *255*: 163–198

Uccelli L, see Boschi A (2005) *252*: 85–115

Vezin H, see Dias N (2005) *253*: 89–108
Vögtle F, see Fages F (2005) *256*: 77–131
Vögtle M, see Žinić M (2005) *256*: 39–76

Warman JM, see Grozema FC (2005) *257*: 135–164
Wasielewski MR, see Weiss EA (2005) *257*: 103–133
Weiss EA, Wasielewski MR, Ratner MA (2005) Molecules as Wires: Molecule-Assisted Movement of Charge and Energy. *257*: 103–133
Weissbuch I, Leiserowitz L, Lahav M (2005) Stochastic "Mirror Symmetry Breaking" via Self-Assembly, Reactivity and Amplification of Chirality: Relevance to Abiotic Conditions. *259*: 177–254
Williams LD (2005) Between Objectivity and Whim: Nucleic Acid Structural Biology. *253*: 77–88
Wong KM-C, see Yam VW-W (2005) *257*: 1–32
Würthner F, see You C-C (2005) *258*: 39–82

Xia W, see Scheffer JR (2005) *254*: 233–262

Yam VW-W, Wong KM-C (2005) Luminescent Molecular Rods – Transition-Metal Alkynyl Complexes. *257*: 1–32
Yoshikawa I, see Araki K (2005) *256*: 133–165
You C-C, Dobrawa R, Saha-Möller CR, Würthner F (2005) Metallosupramolecular Dye Assemblies. *258*: 39–82

Žinić M, see Fages F (2005) *256*: 77–131
Žinić M, Vögtle F, Fages F (2005) Cholesterol-Based Gelators. *256*: 39–76
Zlatušková P, see Stibor I (2005) *255*: 31–63

Subject Index

Adam assembly 126
Adenine 32
Adenosine 45
Adenylate anhydride 101
Adenylate-binding protein 112
Alanine 39
Alcohols 8
N-Alkyl amides 80
Altritol nucleic acid (ANA) 58
Alzulmins 34
Amides 80
Amino acid catalysis 114
Amino acid thioesters 104
Amino acid transport 181
Amino acids, endogenous formation 73
- exogenous formation 79
- extraterrestrial 6
- interstellar medium 82
- non-racemic 81
- phosphoryl 87
- synthesis 69, 73
Amino thioacids 104
Aminoacyl adenylates 94, 100, 111
Aminoacyl phosphate oligonucleotides 103
Aminoacyl phosphoanhydrides 102
4-Aminoimidazole-5-carbonitrile 32
4-Aminoimidazole-5-carboxamide 32
Aminomaleonitrile 32
Aminonitrile 75
- hydrolysis 76
Aminonitriles 104
Anhydrides, mixed 100
Artificial cells 1
Assembly, directed 3
Atmosphere, reducing/nonreducing 74
Autocatalysis, asymmetric 124
Autocatalytic cycles 193
- encapsulation 200

Bags of genes 188
Bilayer membranes 1
Binding energies 90
Biopolymers, self-replication 157
Bücherer-Bergs reaction 77

CAAs (*N*-carbamoyl amino acids) 77
- nitrosation 97
Calvin cycle 195, 199
Carbodiimides 106
Carbonyl sulfide 99
N,N′-Carbonyldiimidazole (CDI) 105
N-Carboxy amino acid anhydrides (NCAs) 88
N-Carboxyanhydrides 89, 147
Catalysis, amino acids/peptides 114
- asymmetric, nonlinear effects 144
Chemomimetic concept 33
Chemoton theory 167, 170
Chirality, clusters 123
- surfaces 123, 128
Cholesterol 8
Circularly polarized light 125
Citric acid cycle, reductive 195, 199
CO_2-catalyzed hydrolysis 94, 96
Coenzyme A-thioester-linked reactions 104
Coevolution, amino acid-nucleotide chemistry 69
Compartmentalisation 186
Cyanate 38
Cyanoacetaldehyde 38
Cyanoacetylene 38
Cyanohydrine 75
Cyanovinylurea 38
Cytosine 38

Dehydration-hydration 14
Diacyldisulfides 98

Dialkylphosphoryl amino acids 87
Diamino acids 81
Diaminomaleonitrile 32, 51
Diglycine 80
Dihydrouracil 39
Diiminosuccinonitrile 51
Diketopiperazine 80, 89
DMPC liposomes 22

Encapsulation 1, 14
- lipid vesicles 14
Enzyme catalysis, primitive 114
Enzymes, adenylate-forming 112
Extraterrestrial infall 5

Fatty acids 8
- vesicles 14
Fischer-Tropsch type synthesis 4
Fluid mosaic model 9
Formaldehyde 43
Formamide 40
Formose cycle 194, 199
Formose reaction 46
Formose/aminothioester 110

Gene redundancy 187
Genomes, primordial 183
Glycine, interstellar clouds 82
- packing arrangement 130
- racemic α-amino acids 140
Guanidine 39
Guanine 33

Handedness, molecular 124
Hexitol nucleic acid (HNA) 58
Homochirality 123
Hydantoins 77, 80
Hydration-dehydration 14
Hydrocarbons, amphiphilic 4
Hydrothermal vents, amino acid formation 78
Hydroxypyrimidine 40
Hypercycles 184
Hypoxanthine 33

Imidazole derivatives 32
Imino diacids 81
Interplanetary dust particles 6
Interstellar ice 82

K_2SO_4 helix, polyacrylic acid 127

Lactams 80
Lattice-controlled polymerization 149
Lipid vesicles 1, 14
Lipid world 168, 173
Lipids 8
Liposomes 22
- PCR 179
Low electron emission diffraction (LEED) 134

Macro-/micronuclei 188
Membrane formation, autocatalytic 173
Membrane permeability 180
Membranes 1
Mercaptopropionic acid 103
Metabolism 194
Meteorites 1, 5
- amino acids 79
Meteoritic organics 5
Methyl mercaptan 16
Micrometeorites 82
Minerals 29
Mineral-water interfaces 15
Mirror symmetry breaking 123, 126
Mixed anhydrides 100
Monoamines 8
Monocarboxylic acids, amphiphilic 1
Montmorillonites 41
Murchison meteorite 17, 37, 79
Myristoleic acid 13

NCAs 88
- energy carriers 111
Nitrosourea 97
Nucleobases 29
Nucleoside-5′-triphosphates 112
Nucleotides 29

Oleic acid 9
Oleoyl anhydride 13
Oligonucleotides, prebiotic synthesis 29, 52
Organo-Zn catalysts 145
Overactivation by cyclization 88, 96
Oxazolidine 77

Parity-violation energy difference 81
Peptide bond stability 84

Peptide cycle, CO-driven 108
Peptide formation 69, 84
- salt-induced 86
Peptide nucleic acids (PNA) 58, 71
Peptides, self-replication 123
Peptidyl-RNA 113
PheNCA, packing 154
Phosphatidic acid 9
Phosphatidylcholine 9
Phospholipids 1, 9
- amino acid polymerization 153
Phosphoric acid, mixed anhydrides 100
Phosphorimidazolide 53
N-Phosphoryl amino acids 87
Photodegradation 81
Plasmids, multicopy 190
Polyalkyl isocyanates 148
Polyethyleneimine 114
Polymerase 22
Polymers, non-folded, enzyme-like 114
Polypentaphosphate 52
Polyphosphate 50
Porphyrin, chirality 137
Post-enzymatic evolution 202
Pre-DNA/pre-RNA 29, 56
Primary pump, dipeptides 107
Protocell model 21
Protocells, growth 13
- sex 190
Protoenzymes 95
Protometabolism 107
Purine nucleobases 32
Pyrimidine nucleobases 37

Quinines, extraterrestrial 6

Racemates, homochiral biopolymers 147
Recombination/gene redundancy 187
Redox state 74
Reducing state, Miller/Oparin 74
Replicase ribozym 172
Replicons 188
Reverse transcription 187
Ribose-cyanamide 46
Ribozyme 21, 23
RNA enzymes 202
RNA polymerase 23
RNA ribozymes 21
RNA world 71, 112, 187

Self-assembly 3, 16
Self-replication, nonenzymatic 157
- peptides 123
Stearic acid 9
Strecker reaction 11
Sugar model, formose/aminothioester 110
Symmetry breaking 123

Template-directed synthesis 157
Thioacids 98
Thiocarbamates 99
Thioester world 104
Thioesters 147
Threofuranosyl oligonucleotides (TNA) 56
Threose 110
Total asymmetric transformation 127
Transmembrane channels 10

Uracil 38

Vesicle division 177
Vesicle models 184

Wetting-drying 15

Xanthine 33

Yamoto-791198 80

Zeolite 40